GREEN INDONESIA
Tropical Forest Encounters

GREEN INDONESIA

TROPICAL FOREST ENCOUNTERS

Written by ILSA SHARP
Photographed by ALAIN COMPOST

Foreword by HRH THE DUKE OF EDINBURGH
Tribute by THOMAS E. LOVEJOY

Research Consultant: ANNE NASH

KUALA LUMPUR
OXFORD UNIVERSITY PRESS
OXFORD SINGAPORE NEW YORK

Oxford University Press

Oxford New York
Athens Auckland Bangkok Bombay
Calcutta Cape Town Dar es Salaam Delhi
Florence Hong Kong Istanbul Karachi
Madras Madrid Melbourne Mexico City
Nairobi Paris Shah Alam Singapore
Taipei Tokyo Toronto
and associated companies in
Berlin Ibadan

Oxford is a trade mark of Oxford University Press

Published in the United States
by Oxford University Press, New York

© Oxford University Press 1994
First published 1994
Fourth impression 1995

British Library Cataloguing in Publication Data
Data available

Library of Congress Cataloging-in-Publication Data

Sharp, Ilsa.
Green Indonesia: tropical forest encounters/Ilsa Sharp:
photographs, Alain Compost.
p.   cm.
Includes bibliographical references and index.
ISBN 967 65 3045 X:
1. Natural history—Indonesia.
2. Rain forest—Indonesia.
3. Rain forest ecology—Indonesia.
4. Natural history—Indonesia—Pictorial works.
5. Rain forests—Indonesia—Pictorial works.
6. Rain forest ecology—Indonesia—Pictorial works.
I. Compost, Alain. II. Title.
QH186.S47 1994
574.5' 2642—dc20
93-46738
CIP

Illustrations on the preliminary pages are as follows:

Printed by Kyodo Printing Co. (S) Pte. Ltd., Singapore
Published by the South-East Asian Publishing Unit,
a division of Penerbit Fajar Bakti Sdn. Bhd.,
under licence from Oxford University Press,
4 Jalan U1/15, Seksyen U1, 40000 Shah Alam,
Selangor Darul Ehsan, Malaysia

*We dedicate this book to wildlife film-maker Dieter Plage,
and to conservationist Ian Craven, who both gave their lives to
the Indonesian rain forest in tragic accidents during 1993,
in Sumatra and in Irian Jaya; and to the many Indonesians
who are toiling often unheralded, for the cause
of conservation in their own country.*

Foreword

INDONESIA carries a heavy and dual responsibility. More than 186 million polyglot and multicultural citizens share their 17,000 islands with a bewildering variety of wild plant and animal species, which inhabit 10 per cent of the world's diminishing rain forests. Providing for the interests of both presents a daunting task for administrators.

It is gratifying to know that there is a growing appreciation within Indonesia of the treasures that lie within its forests and that real efforts are being made to ensure their survival.

The author rightly emphasizes the need to integrate the human community with the natural environment. In Indonesia, as elsewhere on our fragile planet, the future lies in mankind learning to live in harmony with nature.

This book vividly illustrates the range of Indonesia's natural treasures and it gives an indication of the immense diversity of species that exists in this unique chain of islands. I hope that it will encourage conservationists throughout the world to join the people of Indonesia in their struggle to protect these natural riches.

HRH THE DUKE OF EDINBURGH
PRESIDENT
WORLD WIDE FUND FOR NATURE

Tribute

INDONESIA is richly endowed by splendours of Nature. Indeed, the profusion and diversity of the animal and plant life inspired Alfred Russell Wallace to go to the 'Malay Archipelago' to salvage his career after the loss of most of his Amazon collections. In retrospect, it is not surprising that both the science of biogeography and, coincident with Darwin, the theory of natural selection emerged from Wallace's encounters with Nature in what is now Indonesia.

As landmark as Wallace's contributions to natural science are, they represent but an inkling of what Indonesian biology has to offer. The 17,000 islands of this nation represent unduplicable laboratories for the study of life on earth and the advance of the life sciences. With all certainty, that means significant contributions to the life of Indonesians and humanity in general.

Virtually everywhere on earth, the natural world is being assaulted by a tsunami of human population growth and activity which renders insignificant that generated by Krakatau. Indonesia, as a nation with responsibility for a significant fraction of the planet's rain forests and biota, faces a serious challenge to both protect its biological heritage and use its biological resources sustainably. The imperative extends far beyond Indonesia's lands and waters, for the challenge includes global leadership.

The great nations of the decades ahead will surely be those which rise to the environmental challenge we have brought upon ourselves. Only by understanding the beauty, complexity, and variety of Nature can such leadership possibly be achieved. There is no better way to start than to learn from this magnificent volume—and no better way to follow up than to go see Nature in its full glory.

THOMAS E. LOVEJOY
ASSISTANT SECRETARY FOR EXTERNAL AFFAIRS
SMITHSONIAN INSTITUTION
WASHINGTON, DC

CHAIRMAN
TROPICAL FOREST FOUNDATION

Preface

THIS is a book for the layman, anywhere in the world. It is meant to be entertaining, readable, informative, inspiring, and beautiful. Beauty is virtually guaranteed, thanks to veteran wildlife photographer Alain Compost's stunning photographs, the summation of an 18-year labour of love in the wild places of Indonesia.

The popularization of complex subjects like rain forest ecology has its drawbacks and pitfalls, and I apologize for any of these which may be apparent in this book.

The book is, above all, a celebration of the marvels that still reside in the Indonesian rain forests, the stuff of legends and fairy-tales, but real enough for all that.

Equally to be celebrated are the Indonesian government's public commitments, and the Indonesian people's own efforts, towards conserving these treasures, against all odds and despite many obstacles, chief among them being the pressures of a rapidly growing population. This book is therefore written with hope for the future.

ILSA SHARP
Singapore/Australia
January 1994

Acknowledgements

MANY people helped to make this book possible. The many scientists and amateur naturalists in the South-East Asian region deserve special thanks for all the 'training' they have given the author over the years, and for first firing her enthusiasm for the rain forest.

The author would also like to express her thanks to the following:

Alain Compost in Bogor, Java, particularly for his photographs, but also for his patience as we put the jigsaw of this book together;

Anne Nash, formerly in Indonesia and in Malaysia, for her research assistance, text review, advice, and help with picture identification;

Bas E. van Helvoort, conservationist, Bogor, for his meticulous and thought-provoking review of the text;

The Indonesian Forestry Community, especially M. Hasan, Chairman, for access to technical information and enthusiasm for the book;

Malaysian Nature Society, especially Lee Su Win, Executive Officer, for assistance with research resources and picture identification;

Dr Lim Boo Liat, Malaysia, for assistance with picture identification;

Dr Yap Son Kheong, Senior Research Officer, Forest Research Institute of Malaysia, for assistance with picture identification;

World Wide Fund for Nature (WWF) Indonesia, Jakarta, for research materials;

WWF International, Geneva, especially Chng Soh Khoon, Elizabeth Kemf, and Paul Spencer Wachtel, for advice and research materials;

The International Union for Conservation of Nature and Natural Resources (IUCN), Geneva, in particular Chen Hin Keong, Programmes Officer, Asia–Pacific Region, for hospitality and research materials;

Impact (Indo-Mas Pratama Citra) of Jakarta, in particular Shalini Gopalan Menon, for the genesis of the idea for this book, and George Ford and Marianne Pereira, for project co-ordination.

The photographer would like to add his personal thanks to the **Directorate-General of Forest Protection and Nature Conservation (PHPA)**, of the Department of Forestry of the Republic of Indonesia, for all the support and co-operation given to him over the years.

Contents

1 The Turtle Beneath Garuda's Gaze

… in the beginning there was nothing, all was emptiness;
there was only space. Before there were the heavens, there was no
earth, and when there was no earth, there was no sky…. Through meditation,
the world serpent Antaboga created the turtle Bedawang, on whom lie
coiled two snakes, as the foundation of the World.

(The Balinese Creation story from the ancient text, *Tjatur Yoga*,
translated by Miguel Covarrubias, in *Island of Bali*, 1937)

Originally he [Garuda] was the sun itself, which was conceived
by some early Asians as a bird flying across the sky. Today he represents the
heavens…. Garuda is the polar opposite of the earthbound nagas [snakes],
which represent the life-giving terrestrial waters.

(Jeffrey A. McNeely and Paul Spencer Wachtel, explaining the
significance to Indonesians of Garuda, the mystical
golden sunbird, in *Soul of the Tiger*, 1988)

INDONESIA is custodian of one of the world's few last safehouses in which resides Man's most secret inner self. About half of this 5200-kilometre-long chain of 17,000 islands straddling the Equator in the heart of South-East Asia is clad in verdant tropical rain forest, representing some 10 per cent of the world's total rain forest stock, and 40 per cent of all the forests in Asia.

However, the rain forest is more than a mere statistic. It is the mirror into which Man may gaze to see himself and his past. It is where he reunites with his ancient alter ego. To remove totally the rain forest would be to amputate his soul.

During the Pleistocene Epoch of prehistory (2.5 million–10,000 years ago), Man's Palaeolithic or Ancient Stone Age ancestor, *Homo erectus*, also known as *Pithecanthropus erectus*, or Java Man, may have roamed a literally united Asia, with land bridges where today there are great seas dividing land masses into islands. It had not always been so, for up to about 15 million years ago, Indonesia was but a seething cauldron of shifting, grinding geological plates. The country was forged in fire, with violent volcanic eruptions accounting for many of its islands. Some 400 volcanoes, active and inactive, are still visible, standing witness to those times. Still others lie hidden beneath apparently flat land, or below the sea.

At different times, the sea slowly rose or fell, first obliterating, then creating land-forms. Today, the seas have long since risen to make many more or less permanent islands. But the 170 or so active volcanoes that ring the archipelago with fire continue to kiln-create new land, roaring their

◁
Anak Krakatau is the island 'baby' born of the spectacular 1883 eruption of the Krakatau volcano in the Sunda Straits between Java and Sumatra. Anak Krakatau is itself an active volcano, even today.

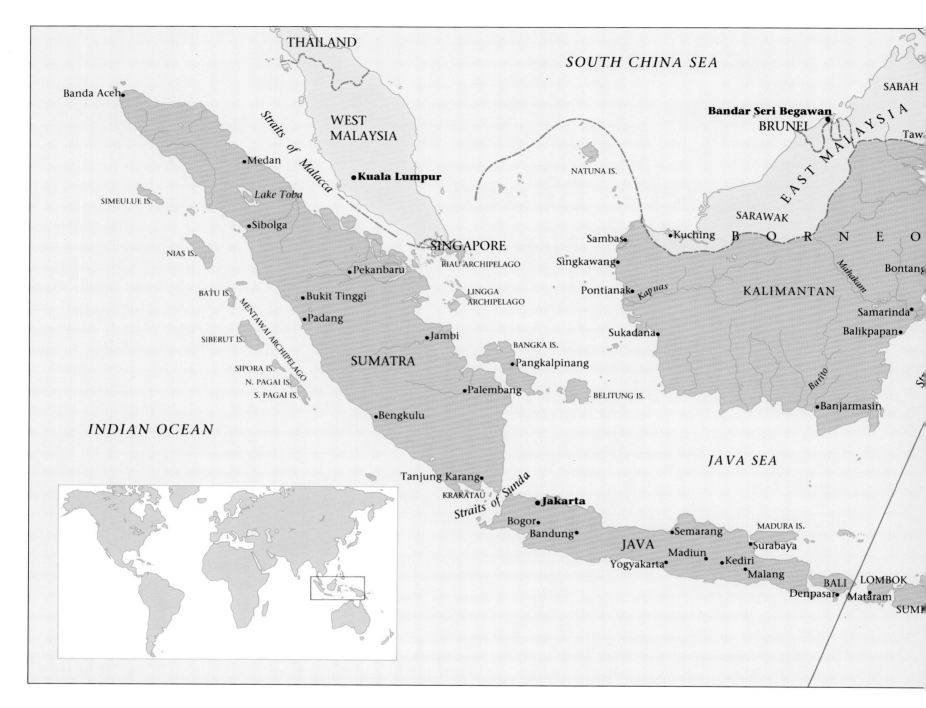

The map shows part of Southeast Asia including Thailand, West Malaysia, East Malaysia, Brunei, Singapore, Sumatra, Borneo (Kalimantan), Java, Bali, and surrounding seas: South China Sea, Indian Ocean, Java Sea.

THAILAND

SOUTH CHINA SEA

SABAH

Banda Aceh

WEST MALAYSIA

Bandar Seri Begawan
BRUNEI

Taw

Medan

Straits of Malacca

Kuala Lumpur

NATUNA IS.

EAST MALAYSIA

SIMEULUE IS.

Lake Toba

SARAWAK

Sibolga

SINGAPORE

Sambas

•Kuching

BORNEO

NIAS IS.

RIAU ARCHIPELAGO

Singkawang

Pekanbaru

Pontianak

Kapuas

KALIMANTAN

Bontang

BATU IS.

MENTAWAI ARCHIPELAGO

Bukit Tinggi

LINGGA ARCHIPELAGO

Samarinda

SIBERUT IS.

Padang

Sukadana

Balikpapan

SIPORA IS.

Jambi

BANGKA IS.

Barito

N. PAGAI IS.

SUMATRA

•Pangkalpinang

S. PAGAI IS.

•Palembang

BELITUNG IS.

•Banjarmasin

INDIAN OCEAN

•Bengkulu

JAVA SEA

Tanjung Karang

KRAKATAU

Straits of Sunda

•Jakarta

MADURA IS.

Bogor•

•Semarang

Bandung•

JAVA

•Surabaya

Madiun

BALI

LOMBOK

Yogyakarta•

•Kediri

•Malang

Denpasar• Mataram

SUM

protest as Australia drifts very slowly northwards towards Indonesia, as if in search of the continental whole that once united the ancient land blocs of Gondwanaland (including Australia) and Laurasia to the north, fragmented and separated 120 million years ago.

The very light of the sun was put out, veiled by something like 1872 cubic kilometres of ash when Mount Toba in Sumatra erupted about 75,000 years ago. South-East Asia's largest lake, Lake Toba, the volcano crater, remains today to help us picture this staggering event. Volcanic activity has continued in this spectacular vein in modern times: in 1815, Mount Tambora on Sumbawa Island to the east of Bali, killed more than 90,000 people, blowing away the top 1250 metres of its own summit, and killing summer in faraway Europe during 1816, causing crop failure and famine across Europe to Ireland. In 1883, the eruption of Krakatau in the Sunda Straits between Java and Sumatra killed more than 36,000, amid huge tidal waves. In both cases, the impact was felt all over the world: ash darkened the skies, enlivening sunsets but cooling world weather for a year; the sound of the eruption echoed far beyond Indonesia. Krakatau still spits and snarls. Every year still sees perhaps a dozen new eruptions country-wide: Mount Agung in Bali last erupted in 1963, killing 1,600 people and leaving 75,000 homeless.

▷
Volcanic eruptions are still a risk in many parts of Indonesia. The 1982 eruption of Mount Galunggung in West Java evicted about 38,000 people from their farmland, and left this wasteland.

INDONESIA

Zamboanga • • Davao

PHILIPPINES

MINDANAO

SULU ARCHIPELAGO

WALLACE'S LINE

TALAUD ISLANDS

SANGIHE
ISLANDS

PACIFIC OCEAN

SCALE

0   100   200   300 km

N

Manado •

• Gorontalo

*Teluk
Tomini*

Palu

• Poso

LAWESI
ELEBES)

BANGGAI IS.

Ternate • HALMAHERA

WAIGEO

SULA
IS.

OBI IS.

SERAM IS.

• Sorong

VOGELKOP

• Manokwari   BIAK

YAPEN

• Windesi

• Waren

Mamberamo

• Jayapura

• Wewak

MISOOL IS.

Fakfak •

• Kendari

BUTUNG

BURU
IS.

• Ambon

MALUKU (MOLUCCAS)

BANDA
IS.

IRIAN JAYA

N E W   G U I N E A

Wanapiri •

**PAPUA
NEW
GUINEA**

ung Pandang
Makassar)

BANDA SEA

KAI ISLANDS

ARU ISLANDS

• Mava

• Kikori

**NUSA TENGGARA
(LESSER SUNDA ISLANDS)**

O

WETAR

ALOR IS.

TANIMBAR ISLANDS

DOLAK

FLORES

Maumere   LOMBLEN IS.   PANTAR IS.   • Dili

BA   **TIMOR**

Merauke •

• Tonda

• Kupang

SAWU IS.   ROTI IS.

ARAFURA SEA

AUSTRALIA

The great Komodo dragon (*Varanus komodoensis*), a carnivore, is a prehistoric relic, and the world's biggest lizard, weighing in at about 50 kilograms and measuring 2.8 metres in length. However, it is not a rain forest species, being restricted chiefly to the open and arid savannah on the islands of Komodo, Padar, and Rinca in the Nusa Tenggara region of Indonesia.

▷

While the population remained small in ancient times, harmony and balance between Man and Nature in Indonesia could be maintained. Today, however, very few Indonesians can live close to Nature in huts like this settler's home on Lombok Island without harming it.

These forests, and Indonesia's rich seas, have provided a living for Indonesian communities for centuries. Today they also form a major part of the resource base for our nation's economic development.

(Hasjrul Harahap, Minister of Forestry, Indonesia, 1992)

Java Man inhabited a magical world of mermaids (dugongs or sea cows, perhaps hippopotamuses) and unicorns (woolly rhinoceroses), giants (prehistoric orang-utans), stegodonts (elephant-like creatures), and dragons (the great lizards of Indonesia's Komodo National Park and giant pangolins or scaly anteaters). Today, Indonesia's rain forests still shelter some of the last descendants of that same magical world.

The non-human denizens of that world betray their fragmented origins: the Asian animals of Java, Sumatra, and Kalimantan derive from the South-East Asian part of ancient Laurasia. The marsupial fauna of Irian Jaya, however, place this Indonesian region of New Guinea island firmly in Gondwanaland. Sulawesi's more ambivalent case has produced a mix of Asian and Australian creatures.

From the beginning in Indonesia, Man was at once part of Nature and yet a threat to it. As do remnant Stone Age peoples in Indonesia even today, those early men lived off Nature's plenty—eating forest fruits and roots, mammals, and birds, using lethal forest toxins to coat their blowpipe arrows and spearheads, and crafting timber, bamboos, and rattans into boats, spears, and containers or fish traps. Java Man had fire and thus could alter the vegetation of his surroundings. But because human populations then were small, there was still harmony and balance.

Slash-and-burn clearance for shifting agriculture of simple root crops was evident in Indonesia as long ago as 12,000 years, at the end of the Mesolithic or Middle Stone Age during which waves of invading *Homo sapiens*, modern Man, began to displace Java Man. It was well entrenched by 5,000 years ago, in the Neolithic or New Stone Age. Wild animals like the red jungle fowl (*Gallus gallus*), the dog, and the pig had also been domesticated by then.

About 2,000 years ago, the 'rice paddy revolution', at first based on simple drainage systems, came to South-East Asia, and soon after, the sophisticated techniques of irrigation agriculture. With these came the need for more systematic human co-operation. Social organization became more complex. The famous Indonesian model for community action at the village level—*gotong royong*—began to evolve. The populations of prosperous rice-farming communities boomed. Cities grew, cultures

This Toraja villager from South Sulawesi clutches a prized rooster which has ancestral links with the wild red jungle fowl (*Gallus gallus*). It is thought that jungle fowls were first domesticated in India, but they were probably already in Man's service by the Neolithic Age in Indonesia.

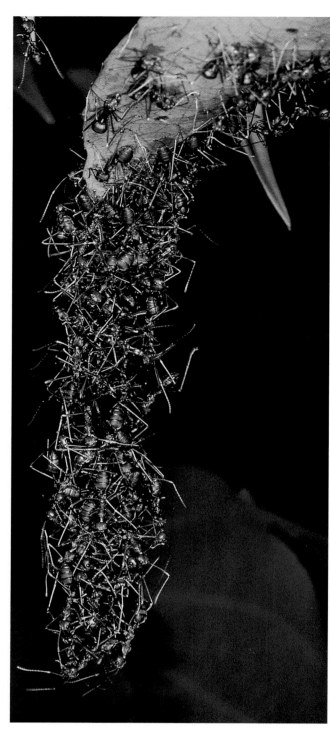

Even an ant like this common weaver ant (*Oecophylla smaragdina*), with its fiery bite, has a right to live, as Islam and many other world religions grant. These ants 'sew' leaves together to form their nest, using the 'silk' exuded from the rear end of ant grubs, which the adults hold like sewing-needles to do the job.

flourished, trade and commerce thrived, empires were built, monuments erected, and kings went to war.

Stone gave way to bronze and then to iron in the first millennium BC, animism and megalithic ('mighty stones') ritual to the great spiritual traditions of India—Hinduism and Buddhism—and of China—Confucianism and Taoism—within the first centuries AD, and finally to those of the Middle East, in the thirteenth to fifteenth centuries—Islam. Christianity, too, was imported, aboard the galleons and clippers of the colonial powers of Europe between the sixteenth and nineteenth centuries.

Each phase left its mark on the Indonesian psyche, but perhaps none truly erased the special Indonesian relationship with the natural world, or with the forest. The gentler, more tolerant and mystical form of Islam which has taken hold of much of Indonesia—Sufism—has nurtured this national trait.

But while it sees them as designed for Man's use and enjoyment, mainstream Islam no less values the animal and plant forms of creation. The scriptures of Islam, collected in the Koran, make it clear that God treasures these creatures not least for their ability to worship and praise Him (in Sura 22 aya 18 and Sura 17 aya 44). An approved prophetic tradition further tells how an ant once stung one of the Prophets, who then ordered the whole colony of ants to be burned in retaliation, whereupon, God rebuked him, saying 'Thou hast destroyed a whole nation that celebrates God's praise, for an ant's stinging thee.'

In the Koran, God says 'There is not an animal that (lives) on the earth, nor a being that flies on its wings, but (forms part of) communities like you (Man).' (Sura 6 aya 38.) Prophetic traditions also tell of the Prophet of Islam himself condemning a woman who had failed to feed or let roam her cat, people who left animals to starve, and people who were found using birds as shooting targets. Islam, then, as do most of the great religions, encourages good stewardship of the earth and all upon it. Approximately 84 per cent of Indonesians profess themselves followers of Islam.

At the very least, closeness to the earth is inherent in the Indonesian character, for even today about 69 per cent of the people work the country's rich soils. Not surprisingly, then, nature reserves are an indigenous rather than an imported concept in Indonesia. As early as AD 684, the great Buddhist–Hindu empire of Srivijaya in southern Sumatra established the country's very first nature reserve.

For a time, mighty Buddhist and Hindu kingdoms were established in Sumatra, Java, and East Kalimantan. But their increasing contact with Muslim Gujarati-Indian and Arab traders later led to their wholesale conversion to Islam. In the late fifteenth to early sixteenth century, the Hindu court of the Majapahit Empire in Java fled the Muslim advance (which was more intellectual than militaristic) to the eastern island of Bali, where the Javanese version of Hinduism has been preserved till today. The rituals of the royal courts at Solo and Yogyakarta in Java likewise retain Hinduistic elements.

Pockets of Buddhism also remain, both in Bali and in remote areas of Java such as the Dieng Plateau. Buddhism and Hinduism have long been entwined in Indonesian culture. The concept of the unity of all life-forms and of the sanctity of life itself, central to Buddhism and Hinduism, and to animism, continues to inform the underlying Indonesian attitude to Nature, giving grounds for optimism about the future of conservation in Indonesia.

There are economic imperatives supporting nature conservation. Forest products made the 'Spice Islands', as Indonesia was once known, a veritable eldorado, luring foreigners to the archipelago. European traders, in particular, braved the extraordinary risks and terrifying unknowns of such long-distance voyages in those days. The courage of the very early European

traders is comparable with that of an astronaut launching into space today, and for the native peoples of Indonesia, these pioneers must at first have seemed much as men from Mars would to us.

The treasure trove sought first by the Chinese in the early centuries AD, also by India, and later by the Arab traders of the thirteenth and fourteenth centuries, comprised spices such as cinnamon, cloves, nutmeg, pepper, and mace, precious metals like gold, cloth, sharksfin, edible swift's nests (of bird's nest soup fame, but at that time used more for medicine), trepang or dried sea-slugs, bird feathers, animal hides, and fragrant woods suitable for incense and perfume, such as sandalwood. In return, the Chinese left silks, porcelain, and bronze wares. The great ships of the Portuguese and Spanish arrived in the fifteenth and sixteenth centuries, but it was the Dutch who were to establish their grip over the key cities and several regions of the archipelago for three and a half centuries, from 1602 onwards, introducing, among other things, extensive plantation agriculture to supply demand in their homeland and other European markets.

Some of the more thoughtful colonials saw that the real treasures of Indonesia lay not in commodity crops but in the country's rich storehouse of natural history. Men like Sir Stamford Raffles of the British East India Company, the founder of modern Singapore and one-time Governor of Java and of Bençoolen in Sumatra, and the English naturalist Alfred Russel Wallace, both made invaluable contributions to nineteenth-century scientific knowledge of Indonesian flora and fauna. Before them there had been Rumphius, the German biologist of the eighteenth century. Many followed, particularly from Holland, notably the great botanist Dr C. G. G. J. van Steenis in recent times, as well as from Sweden and Germany.

A mosque on Madura Island, off East Java. About 84 per cent of Indonesians profess Islam, giving Indonesia the world's largest concentration of Muslims, but within the classification 'Muslim' are contained many shades of orthodoxy.

Nutmeg trees (*Myristica fragrans*) bear the seed called 'nutmeg' and the red seed-skin known as 'mace'. Nutmeg was among the spices sought by early traders as eagerly as gold. Until the early nineteenth century, the trees could only be found on Indonesia's remote Banda Islands in the Moluccas.

Encouraged by the relative prosperity that the largely fertile earth has brought, by the lucrative trading networks initiated by foreigners from both East and West, and by the advent of modern medicine, over 186 million *Homo sapiens* now crowd the Indonesian archipelago, a figure projected to rise to 216 million by the early years of the twenty-first century. But the population is unevenly distributed; it is the cities of Java that carry the main burden, while the wildernesses of Kalimantan and Irian Jaya still contain much uninhabited or sparsely settled territory.

On the island of Java alone, the population has increased 30-fold since 1800, and density there now stands at more than 826 people per square kilometre. National population density is much less, at about 95 people per square kilometre. The national natural population increase is close to 2 per cent a year, and even this represents some achievement on the part of the country's family planners. Indonesia is now the world's fourth most populous nation. Inevitably, there has been competition for land and resources between people and animals, people and the forest, and sometimes between different peoples, too: between the urban and agricultural civilizations of the coasts and valleys, and the older peoples, the hunter-gatherers and the shifting cultivators mostly inhabiting the mountains and forests of the interior.

A settler's *ladang* or plantation in Kalimantan. Indonesia's population pressures seem to make it inevitable that humans should compete for land with the forest and with animals.

Indonesia is one of those few remaining countries where Man still has before him his last chance to reinstate balance with the natural world. He faces a historic choice: to rendezvous with his ancient self and become part of Nature once more, or to leave it forever behind as he voyages on a lonely journey into the unknown. Under the latter scenario, his ultimate destination may well be extinction.

The racial and cultural diversity of the Indonesian people neatly matches the ecological diversity of their resource-rich land. The faces of Indonesia are pale, golden, brown, and black, Aryan, Mongoloid, Malay, Melanesian, Austronesian ... and that is to simplify the import of perhaps 336 different ethnic groups. Linked by a common national language of fairly recent origin—Bahasa Indonesia—the peoples of Indonesia nevertheless speak a total of about 250 languages, not including dialect forms. Yet there is some truth in the maxim-cum-mantra emblazoned in Sanskrit on the national crest and clutched between the sacred Garuda bird's talons: 'Unity in Diversity'. Muslim mosque minarets, Christian church spires, Buddhist pagodas, and Hindu *gopuram* dramatize the architectural skyline of almost every Indonesian town, as do the satellite dishes of the information technology age. Yet in the countryside and remote hinterland, the old animism, megalithic ritual, and even Neolithic lifestyles still survive.

Asmat tribesman, Irian Jaya.

Old man, Madura Island, East Java.

Balinese woman.

Kenyah woman, East Kalimantan.

Toraja man with buffalo, South Sulawesi.

Passion-fruit vendor, Sumatra.

From moist tropical rain forests in the west and north, via deciduous monsoonal forests, to the arid savannah grasslands of the Lesser Sundas in the east, through mangrove swamps and back to the rain forests of Irian Jaya, the 27 provinces of Indonesia reflect the alternation of a hot humid season from May to October with a torrentially wet season from November to April in the west, with dry seasons in the east sometimes as much as seven months long. With such a range of habitats to offer, it is hardly surprising that Indonesia is also a showcase for a huge variety of animals and plants, ranging from Asian to Australian-type species. Many are endemic to Indonesia; they are found nowhere else in the world. 'Cornucopia' might be a better word than 'showcase'. Indonesia has more species of birds and trees than the whole of Africa.

The country is home to about 1,500 bird species (one-sixth of the world's total), more than 380 of them endemic; more than 500 mammal species (about one-eighth of the world's total), 185 of which are endemic; a quarter of all fish (almost 7,000 species); nearly one-sixth of the world's reptiles and amphibians (1,000 species); the largest expanse of the unique mangrove ecosystem in South-East Asia (in Irian Jaya), more than 25 000 square kilometres of mangrove; and more than 10,000 tree species, including hundreds of endemics, or a total of almost 30,000 plant species. One-tenth of the world's flowering plants can also be found in the Indonesian

Rugged hill forest in Sulawesi.

'treasure-house', as ecologist Kathy MacKinnon has justifiably tagged the country. Rain forest occurs in five major regions of Indonesia: the islands of Sumatra, Java, and Sulawesi, and the Indonesian parts of the islands of Borneo (Kalimantan) and New Guinea (Irian Jaya).

Named after Wallace, who holds joint honours with Charles Darwin as progenitor of the Theory of Evolution, 'Wallace's Line' falls between the islands of Bali and Lombok, as well as between Borneo and Sulawesi. In layman's terms, the Line separates Asian- from Australian-type life-forms, but scientifically speaking, it simply denotes a transition zone, a gradual passing from Asia to Australia. Never a clear-cut marker, the Line applies much more effectively to animals than to plants.

Another scientific concept which may bewilder the layman should perhaps be touched upon: that of 'Malesia'. To botanists, in particular, Indonesia is but a part of a larger ecological entity—Malesia—extending from the Malay Peninsula, across Borneo (Kalimantan), taking in the Philippines to the north, and through all the Indonesian islands to New Guinea (Irian Jaya), right out to the Bismarck Archipelago north of New Guinea. With no political or cultural underpinning, this concept resides exclusively in the scientific mind.

Many rain forest species remain undiscovered, especially plants, insects, and almost microscopic organisms like mites. Any or all of them may prove

If we look at a globe or a map of the eastern hemisphere, we shall perceive between Asia and Australia a number of large and small islands, forming a connected group distinct from those great masses of land, and having little connection with either of them. Situated upon the equator, and bathed by the tepid water of the great tropical oceans, this region enjoys a climate more uniformly hot and moist than almost any other part of the globe, and teems with natural productions which are elsewhere unknown. The richest of fruits and the most precious of spices are here indigenous. It produces the giant flowers of the Rafflesia, the great green-winged Ornithoptera (princes among the butterfly tribes), the man-like orang-utan, and the gorgeous birds of paradise.

(Alfred Russel Wallace, *The Malay Archipelago*, 1869)

precious to mankind, whether as medicines or as products of economic value, but it is the value of their survival to the human spirit that really counts. Fortunately, the Indonesian government understands this and has already created 24 major national parks, in addition to the more than 340 nature reserves or protected areas throughout the country, including marine parks. Added to this is an even larger area of forest land which is closed to commercial use. In total, about 25 per cent of the country's land, or almost 500 000 square kilometres, is now under protection.

Despite the demographic and economic factors which might have tempted them otherwise since they achieved Independence in 1945, the Indonesian people have preserved the more than 100 nature reserves established by the Dutch colonials. Many new ones have been added since the Third International Decennial National Parks Conference was held on Bali in 1982, the first time this important assembly had ever convened in a developing nation.

It was the Dutch who made Bogor in West Java and its world-famous botanic gardens, the largest such gardens in the southern hemisphere, the headquarters for research into the natural history of Indonesia. From their inception in 1817, the gardens were the custodian of all Indonesia's reserves. The lush, mist-swathed montane forests of Cibodas in the Puncak Pass close to Bogor were targeted as Bogor's first nature reserve in 1889. Another major park established by the Dutch is Ujung Kulon at Java's south-westernmost tip, today the last retreat for the endangered Javan rhinoceros (*Rhinoceros sondaicus*).

Since 1971, responsibility for nature reserves and parks has rested with the Directorate-General for the Protection of Forests and Conservation of Nature, commonly known by its Indonesian abbreviation, PHPA, to naturalists working in Indonesia. Equally familiar are the formerly joint Ministries for Population and the Environment, also relatively recently

At a bird market in Jakarta, Java. Despite such markets, the Indonesian psyche finds strong affinity with birds, seen as envoys of the soul in the mystic Sufi tradition of Islam.

established, and headed at its inception by widely respected Minister Emil Salim. That the government has taken a personal interest in conservation is understandable, even in purely political terms, and certainly in the economic context.

The Indonesian phrase for 'Our Motherland' is 'Tanah Air Kita', which can be translated more literally as 'Our Earth and Water'. This term, found in the national anthem, captures the typical Indonesian's deep emotional attachment to the physical landscape. The nationally revered Pancasila philosophy which has underpinned the country's sense of social purpose since it was first expounded in 1945 by the charismatic first President, Sukarno, enshrines five basic principles of belief:

Belief in the One and Only God, and with it, belief in the hereafter;

Just and Civilized Humanity, symbolized by a chain of interlocking links;

The Unity of Indonesia, symbolized by a banyan tree, and to be understood in conjunction with the national coat of arms enshrining the principle, 'Unity in Diversity';

Democracy Guided by the Inner Wisdom of the Deliberations of Representatives, based on the old village system of consensus-based decision-making, and symbolized by the head of a banteng cattle; and

Social Justice for all the Indonesian People, the sharing of economic and other benefits and welfare nation-wide, symbolized by a rice-ear and a fruiting cotton twig.

It surely would not have been out of place or out of tune with the thinking of the people had a sixth principle been added—Respect for Nature. Perhaps this addition might yet come.

World-wide, conservationists are increasingly aware that there is no longer any such thing as unspoilt wilderness untouched by Man. Nature can no longer be left alone; it will have to be Man-managed. In Indonesia, too, this realization is dawning.

The fact that there already are spontaneous, home-grown non-government conservation groups in Indonesia, mostly of a 1980s genesis, again supports optimism for the future of the country's unique ecosystems. These groups include the Yayasan Indonesia Hijau or Green Indonesia Foundation, WAHLI (the Indonesian Environmental Forum, attached to the Ministry of the Environment), and SKEPHI (the Indonesian NGO Network for Forest Conservation), not to mention about 956 university, province-, or village-based project groups. The Indonesian government's co-operation with the World Wide Fund for Nature (WWF), Birdlife International, and the Asian Wetlands Bureau on some 130 conservation projects, both completed and ongoing, is also significant.

The national empathy with animals and plants is most visible in Indonesian art and culture, for example, in the motifs adorning the spectacular textiles of the various provinces. Recently too, every Indonesian province adopted an animal and a plant as part of its official crest; in Jakarta's case, the Brahminy kite (*Haliastur indus*), a bird of prey, was the chosen animal emblem. While those who have viewed the caged captives of Jakarta's *pasar burung* or bird markets may find it hard to believe, most striking of all is the natural Indonesian affinity with birds, finding its strongest expression in the national reverence for Garuda. This is a feeling deeply rooted in the mysticism of Sufism, deriving from ecstatic ancient Islamic Arabic and Persian literature which tells of soul birds which unite with the Divine Being.

In many ways then, Indonesians can boast the unique combination of priceless natural assets with the psychological and cultural make-up that could make them the world's leading conservationists. As they approach the twenty-first century, they have become more aware than ever that they carry this burden of responsibility much as the mythical turtle carries the world on his back.

Indonesia's progress is rooted in its human resources and the riches of the land. It is therefore appropriate that we pursue a management strategy combining industrialization with a high regard for the sustainable use of natural resources and allowing precious wild regions to be safeguarded on the path towards a successful future.

(Emil Salim, former Minister of State for Population and the Environment, Indonesia)

How shall I define what thing I am?
Sometimes a mote on the disc of the sun;
At others, a ripple on the water's surface.
Now I fly about in the wind of association;
Now I am a bird of the incorporeal world.
By the name of ice I also style myself;
Congealed in the winter season am I.
I have enveloped myself in the four elements;
I am the cloud on the face of the sky.
From unity I have come into infinity;
Indeed, nothing existeth, that I am not.

(Sufi poem by Mirza Khan Ansari, illustrating the extent to which the mystic Muslim Indonesian may feel at one with the natural world)

2 The Web of Life

LIKE a complex crochet which can be torn asunder by cutting just one thread, the tropical rain forest is delicately balanced. Without the mighty trees, the soil itself dies, sooner or later, for its fertility and its stability derive from the vegetation alone. Just two years after forest clearance, the soil becomes too exhausted for cultivation, hence the nomadic ways of traditional shifting cultivators.

The tropical rain forest is a marvel that even Man with all his accumulated knowledge has not yet been able to invent, replicate, or replace. Once established, it is a self-renewing, self-repairing machine. All it needs is plentiful sunlight, carbon dioxide by day, oxygen by night, nutrients, and water. Undisturbed, it is everlasting, and has already endured for about 60 million years. The cycle of death, decay, and birth in the rain forest has for aeons given Man cause to reflect on his own life cycle and the

◁
The giant trees hold the key to life in the forest, knitting together a complex web. Without them, the soil would die, and with it plant and animal life. This picture of a mighty forest tree's mossy lateral roots reaching into, and across, the forest floor in its search for nutrients was taken in Irian Jaya.

A Sumatran rain forest interior, as delicately balanced as a spider's web.

meaning of death. A forest which literally fertilizes itself, the rain forest allows nothing to go to waste. Rotting wood and leaves and dead animal carcasses are all converted rapidly into nutrients for the living, in a seamless process which cannot be disturbed even by heavy rainfall washing the soil. Chief among the decomposing agents assisting rain forest trees to grow is fungus which coexists with the tree roots, often living inside these roots, in a symbiotic relationship known as mycorrhiza or 'fungus-roots'. The fungus processes dead organic matter into nutrients for the tree, in exchange for sugars from the tree. It is the delicacy of this relationship that is near impossible for Man to restore once his activities have let light into the rain forest, which can destroy the fungus.

Many other agents, such as termites, scavenging beetles, and millipedes, are also involved in this decomposition process; 'saprophytes' is the scientific tag given to plants or bacteria which play this role. Though the forest may at times seem sombre and silent—apart from the general background hum, buzz, and whistle of various insects—it is, in fact, as active as all the offices in a city skyscraper put together. Other agents are also busy generating new life through their food-chains: bats, bees, birds, even elephants, tapirs, pigs, and monkeys act as pollinators or seed distributors when they sip nectar, nibble seeds, or eat fruit.

The forest even waters itself. The tall trees act as a water catchment, attracting rain to the cool air above the canopy, and run it off to feed the plants and soils below.

A 1982 survey by the Food and Agriculture Organization of the United Nations (FAO) estimated that tropical moist forests covered 1.2 billion hectares, or about 7.7 per cent, of the earth's land surface. Of these, well over half were in Latin America, almost one-fifth in Africa, and more than one-quarter in South-East Asia. More than 80 per cent of Indonesia's tropical rain forest is found in Sumatra, Kalimantan, and Irian Jaya. The rain forests of southern Thailand, the Malay Peninsula in Malaysia, Borneo (Kalimantan), Sumatra, and western Java in Indonesia are broadly similar. The tree species in the eastern parts of Indonesia are different, but virgin tropical rain forests of the lowlands generally share a basic structure and certain classic characteristics.

Most striking among these characteristics is biological diversity: the enormous variety of living creatures and plants to be found in such forests is bewildering. Though they cover only about 7.7 per cent of the world's land surface, tropical rain forests hold at least half of the world's species. Sometimes, hundreds of different tree species may be found within a few

A cup fungus (*Cookeina tricholoma*), with a visiting insect inside its cup, at Mount Leuser National Park, Sumatra.

A termite mound, seen in Sumatra. These amazing monuments, which may be as much as 3 metres or more in height, also extend deep underground, honeycombed with complex galleries and chambers. Some species of termite 'farm' fungi to help them make extra food.

Fungi perform a vital role in the cycle of life, death, decomposition, and nutrient-supply within the rain forest. They are able to process tough substances in wood and leaves, such as the polymer, Lignin, and the carbohydrate, Cellulose.

Scientists have identified and named about 1.4 million species of living organisms. Of these around 1.03 million are animals, and 248,000 are higher plants. Our knowledge is very narrow, however; the best studies and most completely known groups are birds and mammals (9,000 and 4,000 species respectively) which together account for less than one per cent of all known species.

(WWF International, *The Importance of Biological Diversity*, c.1990)

square kilometres of what the layman likes to call 'jungle'. (Academic botanists do not like the popular usage of this word, of Hindi and Sanskrit origin, which is more usually applied to secondary (post-clearance regrowth) vegetation than to unspoilt primary forest.)

Sad to say, the rain forest as most of us experience it today is rarely virgin or undisturbed. Hence, at first layman's glance, it often lacks the drama of the African savannah, where huge herds of animals graze on open plains, easily accessible to vehicles and cameras. Such forest may well be old, but its vegetation is still essentially regrowth following human or other disturbance, no matter how ancient. In the shade of true rain forest, visibility is vastly reduced. Animals as big as the elephant may be standing a few metres away from the trekker, and yet still not be seen. It takes some expertise to recognize birds in these forests, too, for they are more often heard than seen.

Transportation necessarily is by *berjalan kaki*, to use the Indonesian phrase for 'on foot', and the process can be quite gruelling, on rough

The warmth and steaming humidity of the rain forest can be sensed in this picture taken at the edge of the Cigenter River in Java's Ujung Kulon National Park.

ground in conditions of high humidity, tropical rain storms, and flash floods, not to mention blood-sucking leeches, thorned palms like the rattan, and poisonous plants like the *Dendrocnide* with its nettle-like stinging hairs which can cause human lymph glands to swell painfully. This may not be conducive to commercial safari tourism, perhaps, but none the less, the true character of the tropical rain forest is very far indeed from the sensationalist image of a 'Green Hell' purveyed in the West for generations, by medieval tall-tale tellers, nineteenth-century colonial explorers and 'great white hunter' yarn-spinners, and by quite a few more recent reincarnations of both types.

Contrary to popular opinion, snakes and tigers do not lurk around every corner of the forest to bother men. In fact, most naturalists consider themselves fortunate if they so much as sight the back end of a tiger or snake running away. Furthermore, the forest is not unbearably hot, but rather cool and shady, and though humid, it is not impenetrable, but relatively easy to traverse. It is also not abuzz with mosquitoes and foul diseases, which are far more prevalent in human settlements at the fringes of the forest, or indeed, in the city. However, many a conservationist has been confounded when, on introducing a city-dweller to the rain forest, the latter has turned round and said, 'Is this it? Is that all there is?'

Unfortunately, to the uninformed layman, the rain forest is just a boring sea of green. Fed with exotic Hollywood images of the rain forest, he or she finds the real thing curiously colourless: flowers of any kind, let alone big, brightly coloured ones, are few and far between, and butterflies are more often of the ghostly black-and-white nymph (*Idea* spp.) type. What there is in abundance is green vegetation—profuse, tangled, climbing, and entwined, soaring skywards like the Gothic arches of a great cathedral. Only knowledge can open the cynic's eyes. Some understanding of forest 'architecture' and a real interest in plants are the *sine qua non* for getting to know and love the rain forest. The wildlife comes later, for forest animals present themselves only to those who can appreciate their home enough to tread softly and patiently in it. To those who simply sit still and wait, the rewards are indeed worthwhile.

Rain forest structure is like a good story, and just as enthralling, too; it comes in three distinct parts: a beginning (the forest floor), a middle (the middle storey), and an end (the topmost canopy level of the tall trees). The forest floor in a pristine rain forest is surprisingly free of plants and quite easy to walk around on, owing to the lack of sunlight; as little as 1 per cent of the available light may filter down through the canopy and mid-storey to the floor.

> There is a touch of something distinctly mystic about the forests, which can only be completely understood and appreciated by one who has had an opportunity to observe them on the spot and with his own eyes … the shadows are so deep that what is really green seems black, and a sort of half-light reigns in this unique realm of nature where the great silence is broken by so many peculiarly strange sounds and voices.
>
> (Eric Mjöberg, *Forest Life and Adventures in the Malay Archipelago*, 1930)

One of the nymph butterflies (*Idea* sp.), from Irian Jaya. Its ghostly fluttering through the forest gives an eerie sensation at times.

A liana or woody climber in Sumatra's Mount Leuser National Park snakes its way across the forest, taking advantage of other trees to expose its leaves to sunlight. Such 'ropes', fit for a Tarzan, may stretch for 60 metres or more.

Only herbs (often the only rain forest plants to offer conventionally attractive flowers), small ferns, algae, lichens (which are fungi–algae partnerships), or mosses and ferns, besides shrub, palm and hardwood tree seedlings, parasitic plants like *Rafflesia*, and some small palms inhabit this level. Among the attractive plants to be found on the floor are the gingers, frequently showing off beautiful red or yellow flowers, and the black lily (*Tacca integrifolia*) with its exotic streamers.

Sometimes the mid-storey of true rain forest does give a somewhat tangled impression reminiscent of the 'jungle' fables of old explorers, with all the creepers, vines, and lianas (woody climbers) of Tarzan-fame clinging on to the tall trees in their common search for light. Quite at home in the shady mid-storey are specialist plants, shrubs, and palms adapted to the lack of light. The rattan palm, for instance, simply hooks its thorns into adjacent vegetation to gain a foothold on its long climb towards the sun.

The tallest trees in the forest are the emergents, so called because they pop up or emerge even above the general level of the canopy. But the most celebrated tree family forming the canopy in lowland tropical rain forest, with their cauliflower-like crowns, is the straight-trunked dipterocarps (Dipterocarpaceae), most of them valuable hardwoods sought by the timber industry, such as keruing (*Dipterocarpus* spp.), meranti (*Shorea* spp.), kapur (*Dryobalanops aromatica*), and ramin (*Gonystylus bancanus*). There are perhaps 380 species to be found in the forests of the Malesian region. Once upon a time, perhaps 40 million years ago, they first came to Asia from Africa. Their stronghold in Indonesia is in the west, especially in Kalimantan. These huge trees derive their family name from the Greek phrase meaning 'two-winged' because of their winged (but sometimes more or less than two-winged) fruits which whirr to the ground like mini helicopter propellers. The wings are meant not so much to carry the seeds long distances, but to anchor in the soil as they hit the ground.

Dipterocarps may attain heights of over 70 metres and build trunk diameters of over 2 metres. If a mighty dipterocarp falls, whether through old age or a lightning strike, or to the logger's chainsaw, it leaves a gap in the canopy through which sunlight can reach the mid-storey of the forest and encourage dipterocarp saplings to replace the fallen giant. Some may wait patiently many years for the light that gives them their chance to reach skywards. This replacement process is very slow, hence the sedate development of any dipterocarp forest. Many of these trees take 70–100 years to reach full height, which is one of the reasons it is so hard to reforest economically with dipterocarps. The mightiest of these forest

At a considerable height, the trees branch off and meet, giving a mutual support and forming a canopy so close and thick as almost to exclude the light at mid-day. Each clump forms with the adjacent ones on every side natural, lofty Gothic arches, which, in the deep gloom that surrounds them ... present as grand and awfully romantic a scene as can well be imagined.

(Thomas Stamford Raffles, describing a forest in the hinterland of Java, 1815, in C. E. Wurtzburg, *Raffles of the Eastern Isles*, 1954)

Rain forest canopy seen from the forest floor, in Sumatra. A tall forest tree reaches for the sky to fill a gap in the canopy.

Emergent trees rising above the general level of the rain forest canopy in Sumatra's Mount Leuser National Park. Many of these are legumes, whose fine leaves do not fight the wind, minimizing the tree's risk of being felled by sudden gales.

giants may be 150–300 years old, or even older. But it seems Nature has thought of almost everything. Certain plant species, nicknamed 'forest tramps' by botanists, are specially designed to take advantage of sudden gaps in the canopy: long-dormant seeds, they speedily germinate, shoot up, and complete their whole life cycle, reverting to dormant seeds on the floor once again, just in time as a slower species such as a dipterocarp grows to close the canopy gap. Yet the balance is fine. Too large a gap in the canopy may let light through to the forest floor, usually a completely shaded level, to stimulate a mass of tangled plants, including the ubiquitous resam ferns (*Dicranopteris* spp. and *Gleichenia truncata*). What was rain forest then becomes 'jungle'.

Pod-fruited legumes (Leguminosae), members of the bean family, are the other tree species besides dipterocarps which dominate as emergents above the canopy. The tallest forest tree you will ever see might be one of these: the sialang (*Koompassia excelsa*), which has the potential to rise 80 metres or more above the ground.

Diversity is the hallmark of the rain forest and so dipterocarps and legumes are by no means the only tree families around. You can also expect to find trees of the puffball-blossomed *Eugenia* group or 'genus', for example, whose edible fruits are known to Indonesians as *jambu*. One familiar member of this genus is the clove tree (*Eugenia aromatica*), which provides the unique ingredient in Indonesia's familiar aromatic cigarette, the *kretek*; the cloves are the tree's unopened flower buds, dried.

Wild fruit trees—rambutan (*Nephelium lappaceum*), jackfruit (*Artocarpus heterophyllus*), durian (*Durio zibethinus*), mango (*Mangifera indica*), mangosteen (*Garcinia mangostana*)—also abound, as do fig-trees (*Ficus* spp.). Increasingly rare is the ironwood tree (*Eusideroxylon zwageri*) in Sumatra's and Kalimantan's forests, from the same family as the more aromatic cinnamon (*Cinnamomum iners*) and bay laurel trees (*Laurus* spp.), an extremely slow-growing but very tough timber tree which can reach 20 metres in height. Forming its own impenetrable canopy, ironwood forest is not really conducive to dipterocarp growth.

Rain forest animals and plants tend to adapt to one or other of the three forest storeys, and can be categorized accordingly. Within their rain forest habitat, they avoid competition with each other for food or for shelter. Some even choose their time-slot, bursting into activity at certain times of the day or night. In this way each species fits neatly into its own place in the jigsaw that is the rain forest ecosystem. Ecologists call that place or role their 'niche'. Apes, for instance, are canopy-dwellers and use brachiation, or swinging by their arms, to move through the branches. Other upper-storey dwellers—flying lizards (*Draco* spp.), squirrels, and flying lemurs (*Cynocephalus* spp.)—can glide from tree to tree. The mid-storey forest is home to countless insects among other creatures, often so well disguised as

The delicate flowers of the ironwood tree (*Eusideroxylon zwageri*), an increasingly rare sight. Slow-growing, it is none the less an extremely valuable and durable timber tree.

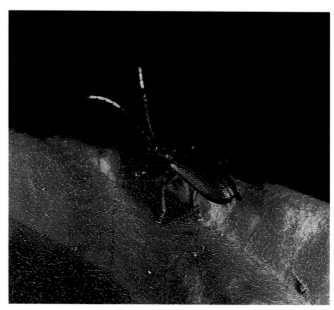

A longhorn beetle (Cerambycidae) makes a handsome splash of colour on a forest leaf.

A flying lizard (*Draco* sp.) in Java demonstrates its ability to glide among the trees. This insectivorous lizard can extend its ribs into struts for flaps of skin which form 'wings'.

to be almost invisible. The tiny cat-sized mouse deer (*Tragulus* spp.), actually not a true deer, roams the forest floor in search of fallen fruit, as well as insects and grubs.

Close inspection of forest trees can yield rewards such as the fissured or flaky bark of a tall *Shorea* dipterocarp and the massive buttress roots of many species. While the forest is mostly green, and evergreen, new leaves are often pink and limp. Another feature of some rain forest leaves is their tapered ends, perfect for rainwater to drip away, thus preventing fungus or algae growth on the leaves. Flowering and fruiting happen at different times, albeit infrequently, depending on the species, probably in response to weather and temperature stimuli. But the dipterocarp family is remarkable for simultaneous mass fruiting at long but irregularly spaced intervals, between five and nine years apart. Rain forest fruiting cycles, therefore, offer wildlife alternating feasts and famine, while the sheer quantity of fruit produced ensures the survival of at least some to disperse and germinate for future new growth. Another feature of many rain forest trees is what scientists call 'cauliflory'—flowering or fruiting directly from the tree-trunk, not from branches—as every South-East Asian devotee of jackfruit or *nangka* knows. This strategy encourages forest-floor animals like deer (and Man perhaps) to nibble on the fruit and so disperse the seeds for the survival of the species.

Sturdy buttress roots support a forest tree in Tanjung
Puting National Park, Kalimantan, and help it search out
the nutrients concentrated on the thin surface layer of
the soil.

Many rain forest plants have leaves specially adapted for
handling the heavy rainfall they experience. The tapered
end of this leaf photographed in Sumatra is perfectly
designed like a small spout, so that the water will run off
the leaf-tip.

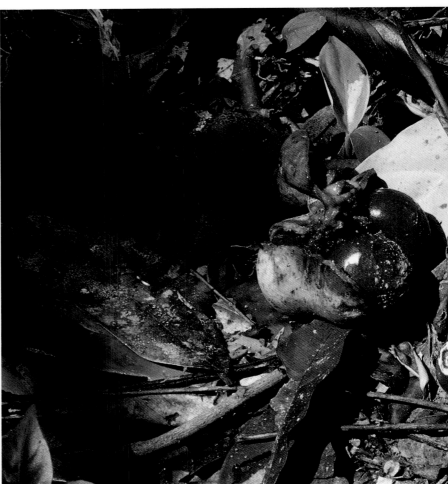

The tokay gecko (*Gekko gecko*), seen here in Java, is a large lizard, up to 0.3 metre long, which makes a harsh cry resembling the sound 'to-kay'. It has strong jaws which can deliver a nasty bite and which enable it to add other lizards, mice, and even birds to its generally insectivorous diet.

Fallen fruit on the forest floor in Sumatra—perfect food for wandering small mammals. These come from a tree in the Meliaceae family, which also contains trees producing fruits favoured by humans, such as the *langsat* (*Lansium domesticum*).

The shrill, rasping 'song' of the cicada (Cicadidae) sometimes approaches the whine of heavy machinery, and is one of the most familiar rain forest sounds. The cicada nymph may spend years underground before emerging to sing in the trees. Pictured here in Java is one of the 2,000 species of cicada, mostly found in the tropics.

While forest sounds mostly reduce to a gentle backdrop bubble of cicadas (Cicadidae) and other insects, verging on complete silence towards midday, there are the occasional dramatic interruptions: the guttural 'to-kay' call of the tokay gecko (*Gekko gecko*), the resonant honk or climactic cackle of a hornbill, the buzz of the montane fire-tufted barbet (*Psilopogon pyrolophus*), the harsh bark of a monkey, the echoing 'ka-wau!' call of the elusive great Argus pheasant (*Argusianus argus*), aptly called the *kawau* in Bahasa Indonesia, or the rasping screech of a cockatoo, and best of all, the joyful whooping and bubbling of gibbons (*Hylobates* spp.) at dawn. Most forest creatures have learned as a matter of safety first to slip through their environment noiselessly, in the very early morning, at twilight, or else at night. Urban Man, by contrast, seems like a lumbering oaf when he ventures into the forest, every brittle leaf crackling beneath his heavy boots, announcing his presence.

Most forest peoples and forest fringe-dwellers have deep respect for the forest and have developed several rituals to request 'permission' for their activities there. These treat the forest very much as a living entity. An old Malay incantation of this type, recited by forest-produce collectors before entering the forest, goes thus:

> Peace unto ye all.
> I come as a friend, not as an enemy.
> I come to seek my living, not to make war.
> May no harm come to me nor mine,
> To my wife, my child or my home,
> Because I intend no harm nor evil.
> I ask that I may come and go in peace.

Most of us no longer have that grace and sense of awe. For the time being, we have lost our original links with the forest. Can Paradise be regained?

A wild jackfruit or *nangka* (*Artocarpus* sp.) in Java exemplifies 'cauliflory', or flowering and fruiting directly from the tree-trunk. The fruit itself is a valuable source of carbohydrate, to humans and animals alike, while its roasted seeds also yield protein.

This montane fire-tufted barbet (*Psilopogon pyrolophus*) from Sumatra is a typical forest bird. This species has a strange cicada-like buzzing song.

What is a Rain Forest?

The scientific world is not yet in full agreement on the definition of a rain forest. However, generally, the rain forest most commonly referred to is usually found in areas of high annual rainfall, in warm, humid regions close to the Equator with little seasonal variation. Such forest comprises tall broad-leaved and evergreen trees, and houses one of the most complex and diverse ecological systems ('ecosystems') in the world, in terms of the myriad species of animals and plants for which it acts as a life support system. Such a forest is also more precisely known as an Equatorial Rain Forest.

The term 'rain forest' was first coined in 1888 by a German botanist, Andreas Franz Wilhelm Schimper (born in Strasbourg, now in France), who travelled extensively in Brazil, Jaya, East Africa, and the Canary Islands. He worked as a professor at the University of Basel in Switzerland from 1898 to 1901. He used the term to describe forests that grow in constantly wet conditions.

Montane forest in Java; also known as mossy forest.

However, the number of different types of rain forest is estimated at 40. Rain forests also occur in temperate regions—chiefly in southern Chile, south-eastern Brazil, Paraguay, Uruguay, south-eastern South Africa, south-eastern Australia, including the Australian island of Tasmania, western New Zealand, the Pacific coast of the USA, and southern Japan. This type of forest is simply known as Temperate Rain Forest.

The main concern of this book, and of most people generally discussing the Rain Forest, is the tall forest found in non-swampy lowland areas close to the Equator. Some attention has been paid too to the higher altitude Montane Rain Forest, which starts as Sub-montane Forest at 1000 metres above sea-level. The sub-montane forest is where you find an abundance of the beautiful tree-fern (*Cyathea* spp.), sometimes mistaken at first glance for a palm. At 1500 metres, the sub-montane forest characteristics accentuate, becoming true montane forest, or what is often called Mossy Forest. At this level, the average annual temperature has gone down by 9–12 °C and rain can be prolonged and heavy. Trees are suddenly much shorter, and strangely twisted as if deformed, almost stunted, as you climb higher.

Mosses and lichens abound in true montane forest, as do myrtles (Myrtaceae)—much larger than their temperate counterparts—and rhododendron shrubs (*Rhododendron* spp.), and tree species typical of northern climes, such as the Asian version of oaks (*Quercus* spp. and *Lithocarpus* spp.). Diversity of species here is rather restricted in comparison with the lowland rain forest. The effect at the higher levels of swirling mist and dripping moss, with gnarled dwarf trees etched in silhouette against the mist, can be quite dramatic and eerie. One is tempted to call it 'elfin forest'. Montane forest can be found all over Indonesia, particularly in Sumatra, Sulawesi, and Irian Jaya.

Monsoon forest on the Aru Islands, Moluccas—a seasonal, deciduous oddity.

Two other rain forest types have been mentioned in this book. The slightly less tall forests found in the tropics within about 10 degrees from the Equator are semi-evergreen and deciduous (i.e. they shed their leaves seasonally). However, since the trees do not synchronize their leaf-shedding, the relevant seasons are almost imperceptible to the casual observer. This type of rain forest, which experiences a brief season of reduced rainfall or even relative drought, is known as Subtropical Rain Forest. The other type of rain forest differs from equatorial rain forest in having a marked dry season, and is known as Monsoon Forest. While the tall dipterocarps can be found in monsoon forest, there are other typical species such as pines (*Pinus* spp.), casuarinas (*Casuarina* spp.), and the much sought-after teak (*Tectona grandis*). Examples of this forest can be found in East Java, Bali, and the Lesser Sundas.

Monsoon and subtropical rain forests are together known as Tropical Seasonal Rain Forest. It is the lowland and montane rain forests combined with tropical seasonal rain forests that some experts refer to with the catch-all tag, 'moist forests', or 'moist tropical forests'. Other types of forest which also technically fall within the classification of 'rain forest' are not dealt with in this book—these omitted forests include mangrove, heath, peat, and freshwater swamp forests, as well as forests on sandy coastal plains, none of which compares in terms of biodiversity with the lowland rain forest. However, the great variety of forest types found in Indonesia includes these categories of rain forest and for this reason, the characteristics of the three most interesting types are briefly outlined below:

Mangrove Forest is miraculously adapted to the oxygen-starved, salty conditions on muddy tidal flats and river estuaries. Among its special features are the strange stilt-roots or prop-roots (in *Rhizophora* spp.) on which it appears to 'walk' above the mud, and spiky roots called

Mangrove forest in Java—miraculously adapted to salt water.

pneumatophores (in *Sonneratia* spp.) jutting out of the mud for all the world like a forest of tiny periscopes. These are the mangrove's 'snorkelling' tubes for breathing. Mangrove trees reproduce through viviparous (live-bearing) seeds (in *Rhizophora* spp.) which germinate while still on the tree, then fall to the mud, rooting there immediately with the help of a spear-like device. These trees can reach heights of more than 20 metres and trunk diameters of more than 1 metre if left alone.

Mangroves are where you can view the magical night spectacle of fireflies lighting up synchronously. They not only host plentiful wildlife, including snakes and the ancient fish-cum-terrestrial creature, the mudskipper (*Periophthalmus* spp.), but also nourish important human food like fish, shellfish, and crustaceans. They are a significant source of thatching or weaving materials and of strong timber too. Unfortunately, as mere 'ugly, smelly swamps', mangroves are the Cinderella of forests and rarely accorded the respect they deserve. Their importance to other ecosystems beyond their own is not yet fully understood but can be guessed at. Yet they are under threat everywhere. Sumatra and Irian Jaya are among the areas in Indonesia still rich in mangrove forest.

Heath Forest is known as *kerangas* (or land too poor for rice-growing once cleared, in the Iban language) in the area where it is at its most extensive—Kalimantan on Borneo. Heath forest grows on poor sand-like soils which are underlaid by a harder, impervious layer offering little or no drainage. The soils are generally acidic and short on nutrients. This type of forest is poor in species variety but very densely vegetated by closely ranked slender trees. The *kerangas* environment is good for specialist plants which have evolved feeding strategies other than nutrition from the soil, like the carnivorous pitcher plant (*Nepenthes* spp.), for example, but generally bad for most wildlife.

Heath forest in Kalimantan—good for specialist plants such as pitcher plants.

Peat swamp forest in Kalimantan—acidic and nutrient-starved.

Peat Swamp Forest is found mainly on the eastern side of Sumatra and in southern Kalimantan. The soils are acidic and nutrient-starved. These peat swamps represent layers of undecomposed plant materials laid down on ancient mangrove soils. Wildlife is scarce in such swamps, apart from some monkeys and crocodiles, but they offer the right conditions for a few dipterocarp trees, for myrtles, and for about 30 palm species, including a wild salak (*Salacca* sp.), related to the rain forest-derived edible salak whose strangely flavoured, crunchy, snake-scaled fruit is so prized at the Indonesian table. The most attractive of these palms is the tall serdang (*Livistona hasseltii*), a feathery fan palm, and the elegant red-stemmed sealing wax palm (*Cyrtostachys lakka*).

A tiny mouse deer (*Tragulus javanicus*), photographed here in Java. One of the rain forest animals that specialize in forest-floor feeding, the mouse deer is a favourite character in Indonesian folk-tales.

Kancil Capers

The mouse deer (*Tragulus* spp.), or chevrotain, found in the forests of Sumatra, Java, and Borneo is not a deer at all. The male mouse deer does not carry antlers, as a real deer would, but has large tusk-like canine teeth on his upper jaw, which he can use for self-defence, much as a stag would use antlers. The barking deer or muntjac (*Muntiacus muntjak*) seems to be a close relative, for it has both the mouse deer's canine tusks and the stag's antlers. Mouse deer are nocturnal and usually solitary, feeding by browsing on shrubs or rooting in the soil for fallen fruit. They very rarely make any noise, but if and when they do, usually a sign of great distress, they make a shrill cry. Breeding mouse deer usually produce only one young.

Combining all the fairy-tale attributes of Brer Rabbit and Reynard the Fox, and not a little of Mickey Mouse, the tiny mouse deer is the hero of many a folk-tale, mostly among the Malay and Dayak peoples. They call him Kancil (pronounced 'Kanchil'), and despite his standing all of 20 centimetres tall, they hail him as 'lord of the forest' for his outstanding cunning.

There is the story of how the mouse deer got the elephants to play football but gave them all broken ankles in the process. Then there is the one where the mouse deer gets an unwilling female sun bear to have sex with him, not to mention the story of how the mouse deer persuaded a raging tiger that if he killed a mouse deer, he would remove too much beauty from the forest.

A favourite Kancil story tells how the mouse deer outwitted the crocodile. Puzzling how to get across a croc-infested forest stream, Kancil engaged the croc king in conversation. The king eyed Kancil hungrily and suggested he come closer, but Kancil was not so easily tricked. Instead, he flattered the croc:

'Oh my! You must have thousands of subjects to serve you,' said Kancil.

'More than you can imagine,' said the proud croc king.

'But I'm sure there are more mouse deer in the forest than crocs, all the same,' ventured Kancil.

Goaded, the croc king roared with laughter and at once summoned thousands of crocodiles to prove Kancil wrong. But Kancil insisted that this was no proof until the king's subject crocs had been counted. Completely distracted, the king agreed to let Kancil count them, and lined them up for him in a nice straight row spanning the river just like a bridge. Kancil leaped from one croc's back to another, shouting, 'One!', 'Two!', and so on, until he reached the other side of the river. Safe at last, he then turned and said to the king, 'You're right! There are more crocs than mouse deer! But thanks for the bridge, anyway!' and scampered off into the forest before the croc king could understand what a trick Kancil had played on him.

The mouse deer is small but not defenceless. Instead of the antlers common to true deer, it has developed long, curved canine teeth for defence.

Togetherness

Great fleas have little fleas upon their backs to bite 'em,
And little fleas have lesser fleas, and so ad infinitum.

(Augustus de Morgan, 1806–1871)

Scientists have a word, or two or three, for it: symbiosis, parasitism, an epiphytic relationship.... We can call it simply, partnership, be it voluntary or involuntary. The whole rain forest is an interactive, interdependent whole comprising many interlocking parts. But some relationships are so close that they warrant special mention. Symbiosis is a 'you scratch my back, I'll scratch yours' relationship of interdependence from which both partners benefit. To use the Indonesian term, it is a form of *gotong royong*. Could the shrew found by our photographer inside the mighty forest flower *Rafflesia* be enjoying such a partnership, perhaps? Or tiny crabs apparently living inside pitcher plants in the south of the Malayan peninsula, a similar environment to Borneo's?

Some of the quaintest symbiotic relationships partner ants and plants. The ant plant (*Myrmecodia* spp. or *Hydnophytum* spp.) is a striking example: inside the plant dwell fighter ants, willing to fend off any outsider tempted to take a nibble at their host plant. The droppings and the remains of their victims left by the ants inside the plant also give the plant an important supply of mineral foods. The ants get a roof over their heads, the plant gets protection and food. Some ants bite off approaching vine or liana tendrils, saving their host plant the extra burden and competition for light that carrying this passenger would bring. Another ant–plant relationship, involving the macaranga plant (*Macaranga* spp.), has the ants favouring their host for nectar-like liquids which the plant has deliberately customized for them alone, to bribe their protection. Or the ants may farm scale insects on the plant which themselves secrete tasty juices, like cattle for milk.

In another symbiotic partnership, lichens, sometimes mistaken for mosses, represent a composite life-form made up of interacting fungi and algae. The fungus controls the shape of the plant and supplies it with water, while the alga manufactures organic food. In this case, the two partners are entirely dependent on each other and cannot live apart. In yet another partnership, trees of the bean or legume family, and dipterocarps, often have bacteria in nodules on their roots which manufacture nitrogen compounds for them, enriching the poor soils around them.

An ant plant (*Myrmecodia* sp.) in Java strikes a deal with its tenant ants: protection in exchange for food.

This lesser tree shrew (*Tupaia minor*) appears to have a relationship with the *Rafflesia arnoldi* flower at whose centre it is pictured, in Sumatra. The shrew may be one of the agents which distribute *Rafflesia* seeds.

A very important partnership is that of the forest plant or tree with its pollinator and seed distributor. These relationships are crucial to the plant's reproduction and survival. When birds or monkeys eat fruit, they scatter the seeds, often through their droppings on the forest floor. An example of the highly specialized relationships that can evolve is the link between the tiny scarlet-headed flowerpecker (*Dicaeum trochileum*) of western Indonesia and *Loranthus*, a mistletoe-like parasite whose glutinous fruits the bird particularly savours. The scarlet-headed flowerpecker has evolved a very short digestive tract specifically adapted for the speedy ejection of the sticky *Loranthus* seeds only minutes after consuming the fruit. The bird then wipes its rear on a branch, giving the parasite its perfect opportunity to start a new life.

Only recently has the role of bats in pollinating the flowers of the mango and durian fruit trees so popular with South-East Asians been fully understood. Squirrels favour acorns, while elephants, tapir, and pigs gorge on fallen fruit on the forest floor. Bees, bats, butterflies, and birds may visit flowers to collect nectar or pollen, but when they do so, they unwittingly carry away pollen dust on their bodies, to be deposited elsewhere, thus fertilizing the flowering plant.

A very specialized relationship is that between various fig-trees and small fig-wasps which pollinate their flowers. Each fig species requires the services of a particular species of fig-wasp, and no other, if the tree is to fruit. Conversely, if the fig-wasp cannot find her own species of fig-tree to lay her eggs in, both she and her eggs will die. The wasp's and the fig-tree's survival are intimately linked. They do not just live together; they have evolved together over time.

Most bizarre of all is the famous parasitic plant *Rafflesia arnoldi*, the world's largest flower. The *Rafflesia*, which is either male or female, emits a foul smell akin to that of rotting meat, but it is none the less attractive to insects which then pollinate the flowers. It is thought that *Rafflesia* seeds may be dispersed on the hoofs of forest pigs or other forest mammals capable of injuring the roots of *Rafflesia*'s unique host vine, to create an opening for the *Rafflesia* seed to lodge in. Similarly repulsive yet attractive is the aroid or lily-like *Amorphophallus* plant, which carries both male and female flowers and also gives out a smell. It cannot fertilize itself and so relies on the visitations of curious flies and beetles to carry its pollen to other *Amorphophallus*. The grotesque *Rafflesia* also offers a good case-study in parasitism. It feeds off the roots of the *Tetrastigma* vine, and that vine alone. More will be said of this oddity in Chapter 4, on Sumatra.

This lichen, seen in Java, is in itself a living partnership, a composite of fungi and algae.

Bees like this one in Java relate to plants, chiefly as pollinators. This bee is feeding off palm fruits.

This epiphytic fern in Java is called the bird's-nest fern (*Asplenium nidus*). Epiphytes simply reside on forest trees, but do them no harm, unlike parasites. This fern collects falling dead leaves in its central cup and thus forms its own soil for its roots to feed on.

In the animal kingdom, of course, parasites, external and internal, abound: all manner of ticks, fleas, leeches, and intestinal worms and bacteria. Epiphytes are often mistaken for parasites by the layman, but are in fact no worse than harmless lodgers hitching a free ride to the sun instead of investing in their own trunk to raise their leaves. An epiphyte will live on another host plant to get light and shelter, but does not damage its host in the process. Some of the most attractive epiphytes in the forest are flowering orchids.

Epiphytic ferns, such as the bird's-nest fern (*Asplenium nidus*) and the staghorn fern (*Platycerium coronarium*) are a very common sight in tropical South-East Asia not only in the rain forest but also in metropolitan parks and gardens. These two ferns and other epiphytic ferns like *Drynaria* simply use their host tree as a support to help them reach for the light. They use their own funnel of spreading leaves to collect the food they need—falling dead plant materials and rainwater. Clearly, Man is not the only species to understand the benefits of co-operation and community.

A Stranglehold on Success

The strangling fig-tree (*Ficus* spp.) is the nearest thing in the plant world to a cuckoo. Worse than a parasite, it is a killer. But like all figs, it acts as an important forest 'restaurant' to a wide range of birds and animals which adore its fruit. A strangling fig's life starts innocently enough, with a fig seed dropped on another tree branch by a munching monkey or bird enjoying a ripe-fig meal, sometimes in the animal's droppings. The fig manages for a while to masquerade as a harmless epiphyte. Then slyly, the seed lowers a dangling aerial root to eventual anchorage on the forest floor where it can feed on various nutrients, and water.

Soon, the full extent of the seed's deceit is revealed. As it grows, the fig drops more roots, which divide, thicken, and fuse into a tough woody latticework, encircling the host tree's trunk. Already, the fig's luxuriantly leafy crown is competing with the host tree's foliage for light. Finally, this tightening, tangled mesh of fig-roots strangles the host tree and kills it. The fig takes over, with the host still rotting away at its hollow centre. Small wonder that, like many fig-trees, this one is regarded by some forest peoples as the home of spirits, to be approached with caution.

The tracery of death, as a strangling fig (*Ficus* sp.), photographed here in Sumatra, grips its victim host tree tightly in its own struggle for life.

Looking up through the hollow centre of a strangling fig, where once the host tree stood before the fig throttled it. This picture was taken in Sulawesi.

A Flash of Colour

Gingers are a joy to the layman naturalist. Any flash of red or yellow spotted on the forest floor, or sometimes growing a little taller than that has a very good chance of turning out to be a member of the Zingiberaceae family—a ginger. These plants are forest-floor herbs. They take many different forms—there are around 1,300 described species—and their flowers vary in colour from brilliant red to a more orange-yellow, or mixtures of the two. Occasionally, a few species of ginger may display delicate white flowers but it is well to steer clear of such species for they are likely to house a particularly feisty little ant.

Ginger flowers may be tiny and trumpet-like, emerging from the waxy lobes of sheathed bracts on an upright stem, often leafless, or spread and star-like, close to the ground. Most last no more than a day. In some species, the flower is tubular and either half or largely buried in the ground. Such species fruit underground, making their method of seed dispersal a bit of a mystery. In most other species of ginger, the seed-box lies at the base of the flower.

From the ginger's long underground horizontal stem or rhizome, which is actually a food larder, shoot a series of vertical leaf-bearing 'stems', an aggregation of leaf-sheaths, with their own root systems. These stems can occasionally reach as high as 3 metres. To quote the late veteran tropical botanist, R. E. Holttum: 'Such a plant as this is potentially immortal. The old parts die after a time, but the growing end can continue indefinitely.' This is one plant which actually enjoys a certain amount of disturbance or clearance in the forest: it often grows at the forest's edge or near clearings, and the presence of gingers in itself is sometimes an indicator of disturbance.

A colourful ground ginger (*Achasma megalocheilos*), found in Java.

Ginger leaves are long and often fairly sturdy; when crushed, they emit the familiar fragrance of ginger. It is, however, the roots of such plants that humans use as spice ginger (*Zingiber officinale*)—*jahe* in Indonesian—and the tasty *laos* (*Alpinia galanga*), or the yellow-staining, slightly bitter-tasting turmeric (*Curcuma domestica*) or *kunyit*, in Indonesian and Malaysian curries. For the thirsty forest trekker, gingers are a boon. The young flowers and half-open flower buds are a juicy, thirst-quenching, if very sour treat, fending off thirst for an hour or more.

Ginger (unidentified species), photographed at Mount Leuser National Park, Sumatra.

Nicolaia solaris, photographed at Mount Gede, Java.

The forest provides its own light. These phosphorescent fungi (*Mycena* sp.) in Kalimantan seem to come straight out of a fairy-tale.

Things That Go Bump in the Night

As evening approaches, one feels more and more the superhuman majesty of the forest. The sea is not more mysterious, more manifestly infinite; but the forest has, beside this, an overmastering personality, so that one feels that one is being constantly watched by some resistless monster, the eternal silence broken only by sounds which seem to intensify it. It is at twilight that one feels the immanence most strongly; and one hardly wonders that the pagan inhabitants should think of the jungle as the home of unknown powers.

(Charles Hose, *The Field-Book of a Jungle-Wallah*, 1929)

Few experiences are more thrilling than treading softly through the velvet black of a rain forest at night, flicking on one's headlamp or flashlight from time to time, to catch a pair of red eyes in its beam—a leopard cat (*Prionailurus bengalensis*), an unblinking owl or a frogmouth bird (*Batrachostomus* spp.), a cuscus (*Phalanger* spp.), a civet (Viverridae), or a mouse deer (*Tragulus* spp.). Nocturnal animals have a reflector at the back of their eyes to give them intensified night vision and it is this reflector that the flashlight catches.

Darkness in the city or even in settled countryside bears no comparison to the blackness of the rain forest at night. Yet flying insects may 'spangle the dark with countless fairy lights', twinkling on and off like shooting stars, as Bornean colonial administrator Charles Hose put it in 1929, while fluorescent fungal strands among the forest-floor leaf litter, and luminous toadstools and fungi can sometimes make the forest floor shine like a carpet of stars. It was the wonder of just such a scene that once caused this writer and her companions to pause in awe in a rain forest and extinguish their lights, only to be rudely jolted from their reverie by the loud warning growl of a tiger near by. Fortunately or unfortunately, only very few night walks get as exciting as this.

The rain forest is as alive at night as it is during the day, but a different shift of animals uses it after sunset. Birds retire from the scene and so do most monkeys and the orang-utan (*Pongo pygmaeus*), but many other mammals prefer the night. Animals such as elephants, pigs, and rhinoceroses are undiscriminating and will forage by day or by night. As the forest closes down for the night, a cacophony of sound erupts, much as it will again at dawn, but without the cheerful whoops of the gibbons' dawn chorus: cicadas whine, crickets vibrate, and grasshoppers hum, frogs squeak, bubble, and croak (many sound like birds), a tiny scops owl (*Otus* sp.) starts its one-note mournful hooting, the tokay gecko (*Gekko gecko*) makes its usual rasping remarks, and the Indian cuckoo (*Cuculus micropterus*) known to Indonesians as *Belanda mabuk* or 'drunken Dutchman', begins its noisy salute to the night.

Bats, of course, are in their element in the dark, and they are usually the first to herald the fall of night. At twilight they can be seen streaming from limestone caves in dark formation trails etched against the greying sky, sometimes hunted as they emerge by a hovering bat hawk (*Macheirhampus alcinus*) who knows their schedule, or, in the case of flying foxes (*Pteropus* spp.) and other fruit bats, pouring off their daytime roost trees. Some flowers open at night specifically for bats or moths to do their pollination chores; the feathery *Barringtonia* in Sumatra is one of these, as is the sought-after durian fruit.

Besides the carnivorous big cats, such as the leopard (*Panthera pardus*) and the tiger (*Panthera tigris*), two delightful primates are also creatures of the night: the saucer-eyed slow loris (*Nycticebus coucang*) and the tarsier (*Tarsius* spp.). The loris (like the mouse deer, it has relatives in Africa) moves painfully slowly in the canopy, using its human-like hands to grab insects, fruits, and shoots or even an egg or nestling when encountered, while the tiny, fist-sized spectral tarsier (*Tarsius spectrum*), weighing only about 100 grams or so, leaps from one vertical tree-trunk to another in the mid- to lower-storey forest, clinging on with its specially adapted finger pads. Sweet-faced though it may seem, the loris is quite happy to eat birds, lizards, and frogs alike, and has very sharp teeth.

The often synchronous flashing of hordes of fireflies bearing lights in their abdomens is one of the still unexplained wonders of Nature to be seen in the forest at night, especially in mangrove. The light is produced by a chemical process in a special organ. This specimen (*Diaphanes* sp.) is seen in Java.

A pair of spotted giant flying squirrels (*Petaurista elegans*), photographed in the Gede–Pangrango region of Java. The species can also be found in Sumatra and Kelantan. Like all flying squirrels, it is nocturnal and can glide long distances by stretching out the loose skin along the sides of its body.

Bats are the only flying mammals and they thrive at night. This short-nosed fruit bat (*Cynopterus sphinx*) seen in Java, like most fruit bats, navigates the night using sight, not the echo-location 'radar' which other species of bat use.

Like the loris, the tarsier also feeds on insects, especially grasshoppers. It marks its territory with urine sprays and also sings in duet with its mate a series of complex calls before retiring for the day, as a warning against intruders. These songs apparently are crucial to the survival and strength of the tarsier couple's relationship and to the successful rearing of their offspring. Family life is of supreme importance to the tarsier: the mother nurses her single young with the greatest of maternal tenderness. The tarsier's widely set huge eyes force it to turn its whole head to look from side to side. This grotesque action has endowed the animal with an eerie, ghostly reputation—hence 'spectral' tarsier—among tribal and forest peoples, who give it a wide berth as a result.

The piebald tapir (*Tapirus indicus*) of Sumatra, a living fossil of 20–50 million years' antiquity related to the horse and the rhinoceros, snuffles the lowland or swamp forest floor for roots and shoots with its short flexible trunk. Although the animal's strange markings are believed to be more of a social recognition signal than night camouflage, these do mimic the shadows cast in the evening forest, and their sharp demarcation breaks up the animal's overall body shape. Tree shrews (Tupaiidae) also probe the leaf litter for grubs, worms, and insects. The disgustingly odiferous moon-rat (*Echinosorex gymnurus*) with its pink nose, is another noisy forest-floor forager, as is the porcupine (*Hystrix brachyura*). On the hunt for termite mounds is the pangolin or scaly anteater (*Manis javanica*), which will roll up into a smooth ball if frightened.

Huge red centipedes (*Scutigera* spp.), poisonous enough to seriously hurt a human bitten by them, and equally toxic scorpions also roam the night forest for insect prey. Orb spiders (Araneidae) spin webs, their clustered eyes sparkling in torchlight, or huntsman spiders simply wait to pounce on their prey. Moths flitter on to their favourite food—rotting fruit. Snakes, too, are active at night—particularly pit vipers (*Trimeresurus* spp.), which feed mainly on rats, kraits (*Bungarus* spp.), coral snakes (*Maticora* spp.), and cat snakes (*Boiga* spp.), which, in turn, feed on other snakes, and the huge reticulated python (*Python reticulatus*) too is out for a kill. They are all happy to avoid the human intruder, sensing the vibrations through the ground from your footsteps. An encounter with a king cobra (*Ophiophagus hannah*) would be an ordeal, however, for when fully erect, this speedy and aggressive snake can stand alarmingly tall—as much as 1 metre.

Man with his limited night vision and his over-fertile imagination feels thoroughly alienated by the nocturnal forest, but Nature has arranged for many other creatures to thrive in this other world of shadows, rustling leaves, and 'things that go bump in the night'.

The aptly named slow loris (*Nycticebus coucang*), seen here in Java enjoys a grasshopper supper. The animal is well adapted to nocturnal life, with saucer eyes to catch available light; it is also suited to its arboreal lifestyle, its hands and feet being endowed with a powerful grip for holding on to branches.

The civet is an omnivorous, largely nocturnal animal, which thrives both at ground level and in the trees. It feeds on small animals as well as fruit. This civet is seen in Java.

Large bifocally adjusted eyes make night-hunting an easy task for this buffy fish owl (*Ketupa ketupu*) in Java. The owl likes to hunt in rivers for frogs, fish, crustaceans, and insects.

Champions of Bluff

Imitation is the sincerest form of flattery, but for some creatures of the forest it is a matter of life or death. To the untutored layman, the forest looks simple—just a mass of leaves. However, you need a 'third eye' to see clearly in the forest. Nothing is what it seems. A green leaf, on closer inspection, may turn out to be a leaf-like insect, a long-horned grasshopper, or a praying mantis; a brown leaf may reveal itself as a moth or even an Asian horned frog (*Megophrys nasuta*), which is so confident of its camouflage that it will sit motionless while you poke your nose up to it and photograph it; some mottled tree bark may transform into a flying lizard (*Draco volans*) or a moth.

That pretty pink-and-white orchid in truth is the orchid mantis (Mantidae), and the bee next to it an orchid. An unremarkable twig is, in fact, a stick insect (Phasmatidae). A white bird-dropping blob is the dung spider (Thomisidae), hiding from spider-eating wasps, while that drop of water on a leaf is actually a tortoise beetle (Chrysomelidae). A patch of foliage dappled by the fragmented forest light suddenly gets up and runs away: the baby tapir is perfectly disguised by its baby hide of yellowish-white spots combined with stripes on darker brown. After about three months, this coat will change to assume the sharply defined black-and-white markings of the adult tapir (*Tapirus indicus*).

Such are the ruses forest creatures use to protect themselves against predators. But if they cannot fool them, then they will try to frighten them, with actual or simulated threats. That dull-looking butterfly will flash open its wings to show fearsome eyes on the upper side. A hawk moth (Sphingidae) caterpillar rears up cobra-like, glaring with its two false eyes, while another moth caterpillar may square up to aggressors much like a large, pugnacious ant. Several spider species make a good pretence of being mini-scorpions, or fiery weaver ants. Beware the fuzzy-haired

The Asian horned frog (*Megophrys nasuta*) can hardly be distinguished from a leaf when it sits still on the forest-floor litter, as seen here in Java. Although these frogs look quite toad-like, and are sometimes called 'toad-frogs', in fact they have the typical moist skin of a frog, not the warty skin of a toad. Projections of skin over the frog's eyes and nose form its 'horns'.

This moth, seen in Java, is mottled to mimic perfectly the tree bark on which it habitually sits.

The orchid mantis, one of 1,800 extraordinary insects in the Mantidae family, hides its predatory purpose behind its flower-like appearance. This specimen is from Sumatra.

The dung spider (Thomisidae) looks like something too unmentionable to attract spider-hunting wasps. Those insects that are attracted are captured and eaten. This spider, shown sitting on its own white silk mat, was photographed in the Moluccas.

The hairs on this moth larva or caterpillar from Java warn 'Danger!' as clearly as the colour red would in human society. Contact with these hairs will cause severe irritation and rash to the human skin, and probably also to the stomach lining of any bird or animal which is foolish enough to attempt eating the caterpillar.

caterpillar, for his ruff is charged with irritant poisons, and think twice, or hold your nose, before handling the colourful stink bug (Hemiptera).

Bright warning colours, such as red, black, white, or yellow, are the warning 'traffic lights' adopted by many forest creatures. Good examples are lacewing butterfly (*Cethosia hypsea*) caterpillars which cluster together to multiply the alarming effect of their coat of many colours, and many luridly marked species of snake too. Other animals deliberately don the masks of more noxious or inedible beasts to ward off their tormentors: longhorn beetles and flies may masquerade as wasps and hornets, cockroaches as wasps, harmless butterflies as poisonous black-and-green striped Danaid butterflies, and crickets as fierce tiger beetles (*Cicindela* spp.).

So perfect is one cricket's mimicry of the tiger beetle that many a professional entomologist has classified it as the tiger beetle. To distract

A tortoise beetle (Chrysomelidae), seen in the Moluccas, successfully imitating a glistening drop of water on a leaf. Enlarged forewings and sides give this beetle its flat but rounded shell-like appearance.

In a few months, this baby tapir (*Tapirus indicus*) from Sumatra will lose its stripes and spots and assume the piebald markings of the adult. Both the baby and the adult are well camouflaged by the fragmented effect of their markings, imitating the dappled light and shade of the forest, especially at night. Surprisingly perhaps, the tapir's nearest relatives are the rhinoceros and the horse.

attention from their heads to less vulnerable body parts, other animals will misplace false eyes at the rear end of their body or on their abdomen. Some butterflies grow filmy 'antennae' at their tail-ends.

It is thought-provoking to compare human behaviour and understand our own need for protective bluff, and for masks of many sorts. It may have started with the warning chest-thumping of the otherwise gentle gorilla, or the hooting and shrieking of the chimpanzee. Face-paint and make-up, menacing tribal (and disco) dancing accompanied by whooping war-cries, dragon-headed Viking warships, and the protective eyes painted on the prows of Chinese sampans in South-East Asian waters, and even our ingenious use of face masks for the back of the head to frighten off Indian tigers, make us as much a part of the animal kingdom as is the orchid mantis or the hawk moth caterpillar.

3 The Tree of Life

JUST as he has always modified his own environment, Man has always been dependent on the forest for important commercial and medicinal products, as well as other more subtle benefits, such as soil husbandry, climate control, and water-supply. The difference today is that the knowledge and understanding of this dependence is not widespread. Ancient wisdom is no longer community property, to the point that many people ignore or even deny Man's relationship with the forest.

Although some 40 per cent of all prescription drugs are derived from natural materials, we conveniently forget the forest origins of rubber, chocolate, kiwi fruit, malaria, blood-pressure, leukaemia, and headache medicines, and of countless other frills of civilization, just as we gloss over the fact that to get a frozen chicken into a supermarket, somebody has to kill it. In industrialized and post-industrial societies, despite advances in education, we have taken a step backwards: we are no longer 'forest-literate'. We no longer understand that forests are 'user-friendly'.

◁
Man and the forest are locked in natural balance. Here in the heart of the forest lie many of the secrets of life for those who have the patience to seek slowly, carefully, and respectfully.

Traditional peoples need forest produce to make attractive and saleable handicrafts, such as these examples from the Dayak and Kenyah peoples of the Mahakam River area in East Kalimantan.

An Asmat warrior in Irian Jaya has strapped to his arm his trusty knife, crafted from the bone of a wild cassowary (*Casuarius* sp.). In his most natural state, Man is dependent on animal and plant resources for all his most basic needs.

Sculpted relief walls at the ancient (eighth- to ninth-century AD) Buddhist temple of Borobudur in Central Java depict the Tree of Life motif which is still at the core of Indonesian culture and its concept of order and balance in Man's relationship with the natural world.

The livelihood of many rural Indonesians depends on clearing the forests to plant rice or another crop, then moving on to clear more land the next year. Outside of street-corner begging, their only other survival option is salaried employment. That is exactly what the forest industry provides: employment for almost four million workers at all skill levels.

(M. Hasan, Chairman, Indonesian Forestry Community)

Indonesia is among the few places left in the world where there are still many simple, traditional, tribal, and rural communities who may not have been to university, but who understand and live Man's relationship with the forest. Such people instantly recognize, and subscribe to, the 'Tree of Life' motif central to their culture, which depicts human society under the shade of a tree alongside animals such as the buffalo and the tiger, locked together in natural balance.

Nobody inspecting Indonesia's trade figures, now or in antiquity, could doubt the importance of the forest and its survival to the nation: to cite the most glaring example, timber or timber-derived earnings were US$5.2 billion in 1992 (about 11 million cubic metres of tropical timber were exported annually in 1989). Timber is now the second most valuable export after oil (surprisingly, only about 30 per cent of the gross national product derives from agriculture). In 1992, the value of Indonesia's timber-product exports accounted for over 15 per cent of all export value.

Exports of non-timber forest produce, such as furniture rattans and resins for industrial materials, incense, or perfume, were worth more than US$224 million to Indonesia in 1988. Reigning supreme among non-timber forest products is the humble rattan palm. It is not often realized that rattan palms—and there are some 300 species in Indonesia—are probably the next most threatened plants in the rain forest after the timber-bearing dipterocarps (Dipterocarpaceae), such is their economic value to human society. Indonesia was the source for almost 90 per cent of the world's rattan trade in the early 1980s, exporting US$97 million worth of unprocessed rattan cane in 1985. Finished rattan products alone were worth about US$248 million in 1992.

Rattans—locally known as *rotan*—are climbing palms. In tropical rain forests, they account for almost two-thirds of all palms. The majority belong to the genus *Calamus*, the next largest group being the *Daemonorops* genus. Forest trekkers are made quickly aware of rattans, thanks to the plants' vicious long thorns arrayed all the way up their stems, and the hooks displayed on whip-like extensions which they use to claw their way up to the canopy light. Nothing could be more daunting than a walk in a rattan forest.

Strong and springy, of various shapes and sizes, and capable of growing to lengths beyond 100 metres, rattans are ideal for furniture manufacture or basket-weaving. They can be used to fashion ropes, walking sticks—the original Malacca cane (*Calamus scipionum*)—fish traps, or suitcases. Some species endowed with a natural shine are particularly suitable for the

Boat-building from forest timber, South Sulawesi. Besides being commercially important, such timbers also serve many traditional Indonesian users.

Rattan products constitute one of Indonesia's biggest revenue-earners in the non-timber forest produce category.

Rattan fruits like these ones pictured in Java are a thorny proposition to harvest but can yield a useful dye.

manufacture of furniture and split-cane mats. Once collected, the rattan is usually hand cleaned, then boiled to remove natural gums and resins, and sometimes split before being bundled for sale.

Indonesians very much enjoy eating the heart of certain rattans—the bunch of folded immature leaves at the centre of the plant. Forest people also know how to extract from the scaly rattan fruits a useful dye, a dark red resinous substance which collects under the fruit, known as 'Dragon's Blood'. These fruits have to be harvested with caution, since many are housed within spiny sheaths, and several species of rattan also harbour martial ants. The relationship between certain bees, wasps, and beetles often found on rattan flowers, and the rattan palms themselves, has yet to be studied.

Rattans are extremely difficult to cultivate outside the rain forest. Thus, only about 10 per cent of Indonesia's total rattan exports originate from plantations, mainly in Kalimantan and Sulawesi. Being climbers, rattans need the tall forest trees for support. Furthermore, their seeds—only one per fruit—can take up to six months to germinate on the forest floor, and are dependent on the complex interactions of beetles, bacteria, fungi, and insects that maintain rich and moist forest soils. Over-dry or bright conditions in open clearings would be a death sentence for the rattan.

Heavy collection is threatening some rattan species, notably the ones with the largest diameters. In a recent WWF report, a manager of a rattan workshop near Kerinci, Sumatra, complained that it was getting harder to obtain the rattan: 'Each year we use 10–20 tons of thin rattan, and 20,000 sticks of thick rattan,' he explained. Reports from Kalimantan suggest that rattan collectors have to travel on foot for about

200 kilometres at a time, and then stay 10–20 days at the rattan harvesting sites to collect effectively. None the less, rattan production rose from 57 651 tonnes a year in 1976, to 109 535 tonnes in 1987. Careful husbandry of rattans both inside and outside the forest promises big returns for Indonesia.

The question of rain forest timber and its harvesting is a large and complex one, which deserves a book in itself, and which has already been discussed at length in many other publications. Suffice it to say that, at the highest levels of Indonesian society, the 1990s have seen a new understanding of the mistakes of the past, in terms of some ill-advised logging practices (although clear-cutting of timber has never officially been permitted), and a new resolve to protect forest and wildlife resources in future: witness the Ministry of Forestry proposals in 1993 for new conservation areas and extensions or rationalizations of the boundaries for existing areas, totalling 110 000 square kilometres, in addition to the then existing 300 000 or so square kilometres of conservation areas.

Since 1980, the Indonesian government has used various methods to discourage the export of unprocessed logs, and since 1979, of raw rattan too, with semi-finished rattan exports prohibited since 1988. It has been encouraging, instead, value-added domestic processing and finished-goods industries. In late 1989, such high taxes were imposed on sawn-wood exports that this trade, too, is expected to phase itself out. The hope is that such measures will ensure the benefits of forest produce earnings and factory employment may at least flow more directly to the people of Indonesia itself.

Thus, Indonesia's forests are important not only to the republic itself, but

Towing logs along the Mahakam River in East Kalimantan. There is now a high tax on the export of unprocessed logs.

◁

The spiny rattan palm (*Daemonorops* spp.) is a major source of export revenue for Indonesia and is used for a wide range of purposes, particularly furniture-making and basket-weaving. Rattan hearts also make good eating. This palm was photographed in Irian Jaya.

Crocodile farms such as this one in Medan, Sumatra, may be an acceptable form of wildlife farming which will help Indonesia earn revenue from its natural resources in the future, without harming conservation programmes.

also to the world as a whole. Apart from their possible impact on world climate and the notorious greenhouse effect, they also represent a storehouse of genetic stock which may be needed at any time to strengthen cultivated strains or domesticated species elsewhere in the world.

With regard to the wildlife, ranch-type operations are particularly attractive ways of redirecting poor Indonesians' cash hunger away from trade in wild caught animals, and of integrating human needs with those of Nature. Not all trade is stimulated by hunger or greed alone, however. To understand the trade, more conservationists need to develop some empathy with the complex emotions underlying it. Some buyers of wild animals will tell you how much they 'love' animals, and curiously enough, the sentiment is often genuine, albeit misplaced. This particularly applies to caged songbirds, undeniably treasured by their owners. The caged songbird is entrenched in cultural traditions within Indonesian society, making it all the more difficult to protect wild birds in the South-East Asian region. The collector's lust for pinned butterflies can also be attributed to a genuine, if perverted, admiration for their beauty. To meet such trade halfway, there are examples of successful ranching operations in Indonesia for several species, but probably the most exotic project is a butterfly farm.

Birdwing butterflies (Papilionidae) are the crown jewels of Indonesia's forests. Their wing-span can reach 33 centimetres, and they are opulently coloured in rich greens, blacks, yellows, oranges, and browns, with occasional touches of red. Some birdwing species live in lowland forest, like the four endemic to the Moluccas; others stay at canopy level in mountainous forests like those of the Arfak Range at the western tip of Irian Jaya. The beauty of these insects makes it hardly surprising that in South-East Asian urban centres like Jakarta, Singapore, and Kuala Lumpur,

A perverted love of beauty—the collector's love for pinned dead butterflies. Such cases are seen in tourist souvenir shops all over South-East Asia. They often include Indonesian species.

A birdwing larva or caterpillar (*Ornithoptera* sp.) in Irian Jaya. Learning more about these creatures' feeding needs may benefit not only would-be butterfly farmers, but also conservation programmes.

One of the Moluccas huge birdwing butterflies (*Ornithoptera croesus*), which often has a wing-span of nearly 20 centimetres. Farming such butterflies for the collector's market may be the answer to giving rural people an environmentally friendly way of making money from Nature.

souvenir shops are lined with presentation cases full of their pinned bodies, a lepidopteran mortuary for sale. A single specimen, especially of the rarer species, may fetch US$100 on the market, or far more alive. World-wide, the butterfly trade is worth about US$100 million a year.

A WWF project in Irian Jaya's Arfak Mountains region, home to six birdwing species, hopes to repeat successes already scored in neighbouring Papua New Guinea (where successful crocodile farms are also pointing the way for Irian Jaya), farming birdwings to feed this market, and at the same time offering the local people some source of income other than cutting the forest. In Papua New Guinea, which has gone so far as to enshrine the principles of butterfly-farming in its constitution, farmers can earn as much as US$1,200 a year from their winged 'flocks', which in a nation with an annual per capita gross domestic product of only around US$717 (1993 figures), is considerable. The official PNG butterfly trading agency sells about US$200,000 worth a year.

The Arfak Mountains region is blessed with hundreds of New Guinea's 5,000 species of butterflies and moths, many of them endemic. The WWF's butterfly-farming technique also protects the butterflies in the wild, since all the Hatam tribal people of the area have to do is plant whatever attracts the butterflies and then collect either their pupae or newly emerged butterflies, leaving many to return to the wild. This is a very acceptable low-technology, non-capital-intensive style of farming, and a good example of buffer zoning to protect natural areas—in this case the Arfak Mountains Nature Reserve—against human intrusions for economic purposes. The farmers need the rain forest to foster butterflies which will come to lay eggs on their plants, thus augmenting their income beyond what they could earn from the back-breaking work of growing vegetables, which would entail rain forest clearance.

As with all compromise solutions, the project is not without its flaws. Its implied acceptance of the burgeoning market in pinned butterflies is controversial, for example. But the Hatam apparently love the idea and are willing to be trained in the handling, packing, and transportation of the butterflies. As WWF's Paul Wachtel remarked, the list of would-be Hatam participants in the scheme has proved to be a more complete census of local inhabitants than the official government count. Such experiments are a necessary phase as we grope our way towards that elusive natural balance between Man and the forest.

Eco-tourism—the marketing of natural attractions to tourists—is yet another option for Indonesia in the search for non-timber forest revenue. The popularity of projects like the Bohorok Orang-utan Rehabilitation Centre in Sumatra, inside Mount Leuser National Park, which attracts about 5,000 domestic and 1,000 foreign visitors every year, is a case in point. While such projects do not offer the visitor forest, the success of the volcanoes and Hindu festivals of Bromo Tengger Semeru in Java is instructive: they lure over 150,000 people a year. Further to the east, the Komodo dragon lizards (*Varanus komodoensis*) demonstrate that there are still other attractions with potential for further development. But eco-tourism is not the easy option it seems, for it is not just a matter of erecting a five-star hotel in the middle of a national park. Besides, the dense, shaded tropical rain forest does not lend itself to African savannah-type accessibility, visibility, or transportation comforts.

The packaging, marketing, and interpreting of a rain forest experience for an international layman audience is therefore a particularly expensive and skilled enterprise. There is always the fear, too, that increased tourism, both domestic and foreign, could kill the goose that lays the golden egg, in terms of the human impact on delicate ecosystems—pollution, noise, the impact of many feet on soils and seedlings, vandalism.

Eco-tourism in Kalimantan. This is not as easy to manage in the rain forest as on the African plains.

Thus, capital- and labour-intensive monitoring and control systems, education and law-enforcement programmes, perhaps also upmarket pricing to discourage excessively high-volume tourism, are essential to success in the eco-tourism endeavour. Privatization is the preferred approach of many South-East Asian governments to eco-tourism development projects, but proper control of such private enterprise may be yet another problem area, as is careful integration of local people's needs and desire for participation in private projects.

Undoubtedly the most exciting potential of the Indonesian forest, still only partially tapped, lies in non-timber produce of the vegetable kind. The value to Indonesia of non-timber or 'minor' forest produce, as it is sometimes dismissively described, increased by at least 198 per cent between 1977 and 1981. Extraction of such materials is the best alternative to timber logging for the future. These materials include not only well-known items such as rattans, but also more obscure products

Harvesting *damar* resin from an agathis tree (*Agathis* sp.) in West Java. The resin is used in the manufacture of lacquers and varnishes.

Sago, here chewed by a child of the Asmat tribe in Irian Jaya, is an important source of edible starch in many parts of Indonesia. It derives from the sago palm (*Metroxylon sagu*). The Asmat also rely on the tree for building and thatching their houses, and for weaving baskets.

such as gaharu, also known as aloes wood, or eagle wood. Gaharu is a rare and precious resinous product found only in certain diseased trees (*Aquilaria* spp.), mostly in Kalimantan, and is traditionally used for treating ailments, especially those connected with pregnancy and childbirth. However, it is also traded internationally for uses such as incense manufacture. In 1980, gaharu was priced at about US$100 per kilogram in Indonesian cities.

A full list of 'minor' non-timber forest products includes wild meat, fish, including valuable aquarium fish like tiger barbs (*Puntius tetrazona*) and rasbora (*Rasbora* spp.), fruit, orchids, edible oils, spices, sandalwood (*Santalum* spp.) and other aromatic timbers, fuelwood, and charcoal; construction materials such as rattan, bamboo (Bambusaceae), of which Indonesia has 35 species, poles such as ironwood (*Eusideroxylon zwageri*) in Kalimantan, and various fibres; chemically useful materials such as resins (*damar* is the widely used Indonesian word for these), camphor (*Cinnamomum camphora*), essential oils, gums, latex, tannins, and dyes; animal products such as honey, eggs, lac (resins secreted by tiny forest lice and used for shellac varnish), silk, reptile skins, feathers, live animals, and ornamental plants.

The more exotic recipes for traditional medicine and magic list many more animal parts. Included among these are rhinoceros horns and tiger bones, hornbill casques, and, most peculiar of all, bezoar stones, found in the gall-bladders of a few species of monkey. The magic powers of bezoar stones are employed by forest shamans to invoke the tiger spirit and exorcize evil spirits from sick people's bodies. Bezoar stones are also used to combat chest and bowel complaints, and as aphrodisiacs.

Some of the more poisonous products, such as those in the *Derris* genus of vines, yield useful insecticides since their roots harbour rotenone, a commercial pesticide. From Indonesia's 477 native palm species alone (225 of them exclusive to Indonesia), leaving aside the commercially important rattan palms, can be derived important edible starches (such as sago and nipa), other foods including fruit, coconut products, sugar, alcohol, coffee substitutes, building and thatching materials, fibres, cords, weaving and handicraft materials, blowpipe dart poisons, charcoal, wrapping and packaging materials, dyes, medicines, and ornamental pot-plants.

Apart from edible wild animals, some of which, like the banteng beef cattle (*Bos javanicus*) or the red jungle fowl (*Gallus gallus*), are also the ancestors of domestic livestock, there are also important vegetarian foods to be harvested in forest or forest-fringe settings, such as the crunchy sweet jelly-like seeds or heady wine from the swamp-loving nipa palm (*Nypa fruticans*), the oil-bearing *Shorea* nuts often called illipe nuts (whose oils substitute for cocoa butter in chocolate manufacture, and lubricate many a lipstick), the kidney-cleansing but strong-smelling *peteh* beans from species of *Parkia* trees, breadfruit (*Artocarpus altilis*), fern tips and yam roots (*Dioscorea* spp.), gingers and mushrooms, and perhaps 100 species of wild fruit trees—and many more as yet unknown, or not described by science. The WWF estimates that 90 per cent of the world's plants are as yet unidentified, while 25 per cent of all plant species are under threat.

Our knowledge of medicinal plants outside of traditional and rural societies is therefore still limited. Malesia-wide (in Indonesia, Malaysia, the Philippines and Papua New Guinea/Irian Jaya), there are about 40,000 plant species to study, many of them endemic, a lot of them medicinal. Scientists

Wild pigs, like this one just dispatched by an Asmat hunter, are not only important food-animals but also central to social and ritual life for many tribes in Irian Jaya.

This wild example of the salak fruit (*Salacca conferta*) from Java is the source for the more domesticated and popular, 'snake-scaled' edible salak (*Salacca zalacca*). Both are palms.

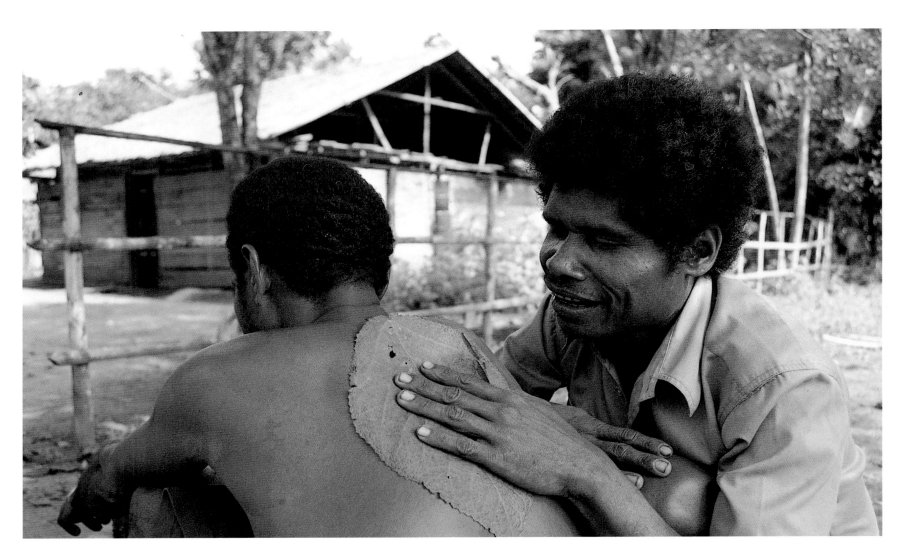

On Seram, in the Moluccas, the people use a certain 'stinging leaf' for various ailments. This is just one of the approximately 7,000 wild plants in the traditional Indonesian pharmacopoeia.

have estimated the number of plants used by traditional healers in East and South-East Asia at over 6,000 species, and Indonesians are known to find at least 7,000 species useful in some way or other. Among these useful plants are 500 species of fruit trees, 340 vegetables, 1,000 medicinal plants, 200 fodder plants, more than 150 edible roots and tubers, and 220 dye and tannin-bearing plants.

A study in North Sumatra's Mount Leuser National Park in the early 1980s recorded more than 200 medicinal remedies available from 171 plants, as used by the Gayo tribal people. On Siberut Island, off western Sumatra, the bark and leaves of a wild fig are used by the local people to combat giardia, an intestinal parasite uncomfortably familiar to travellers in Indonesia. Various roots are favoured in Sumatra and Java for the treatment of malaria, impotence, and unwanted pregnancy, among other human problems. The leaves of certain paperbark or *Melaleuca* trees native to East Java and Irian Jaya, and also cultivated on other Indonesian islands, have traditionally been used to produce cajeput oil (or *kayu putih*, white-wood), for muscular pains and headaches.

The US National Cancer Institute, together with the WWF, recently conducted a five-year research programme in South-East Asia, based on collections of plants from the rain forest with potential for curing cancer. The way ahead here can be glimpsed in Thailand's booming plant-based pharmaceuticals business, worth US$18 million on the domestic market in 1979. Thai indigenous herbal preparations are now supplanting foreign drug imports and therefore saving foreign exchange; in 1986, they accounted for 49 per cent of the sector.

The meteoric success of health-and-beauty ventures such as The Body Shop, which, though British-based, has increasingly used forest-derived herbal concoctions from traditional or tribal societies, points to the growing commercial importance of non-timber forest produce. Another

At present we know more about the surface of the moon than we do about tropical rainforests. Yet what we have learned about them so far has revolutionized our view of all life on this planet.

(Catherine Caulfield, *In the Rainforest*, 1985)

indicator is the hunger of Indonesian women for their rejuvenating beauty and health panacea, *jamu* (equally sought after by the men as *obat kuat* or virility-promoting aphrodisiacs), also largely derived from forest plants and herbs such as gingers.

Here again, conservation will require wise use of such herbal resources, lest over-collection from the wild leads to the extinction of many desirable plants. Fewer than 15 per cent of the plants which Indonesians find useful are actually cultivated; collection is mainly from the wild. Although it would be naïve to imagine that plantations automatically protect the wild forest, the need to domesticate and farm such plants is abundantly clear.

The biggest problem is with the plants which are collected for their roots—or for their bark in the case of trees—and which are therefore automatically killed on harvesting. There are already reports from Java that some medicinal plants are in short supply, for example, the species of *Rauwolfia* whose roots are used to make reserpine-type drugs to tranquillize and to lower blood pressure. *Jamu* practitioners are already having to substitute various species of the cinnamon tree for certain scarce ingredients. The creation of bodies like the National Commission for Conservation of Plant Germplasm recently established by the government of Indonesia is particularly welcome, as is the way the Department of Health has actively encouraged home gardeners in their traditional practice of planting medicinal plants for home use. Some Indonesian botanists have also urged that unproductive *alang-alang* grasslands be planted with such herbs.

Indonesia has a hard road ahead in reconciling the many demands of its people with conservation imperatives, meriting both understanding and, most of all, concrete assistance. The journey has already begun. The goal is simply the restoration of Man's natural harmony with Nature, under the umbrella of the Tree of Life.

The almost mystic cure-all of Indonesia, *jamu*, can be drunk, eaten, or applied. Indonesian men and women swear by it, not least for its alleged properties of rejuvenation. Most *jamu* is derived from forest plants and herbs such as gingers.

4 Sumatra: The Island of Gold

INDONESIA'S westernmost and second largest island at 473 606 square kilometres (including some 'satellite' islands), nudging the Malaysian peninsula and Singapore to the north-east, Sumatra today retains the mysterious allure which won it the tag 'Island of Gold' among Indian traders back in the early centuries AD. The island is still more than 43 per cent forested and still peopled by several extremely diverse and exotic civilizations. These range from the once-warlike, now songster Bataks of the northern Toba highlands, to the megalithic villages of the Niha on Nias Island, the canny, matrilineal Minangkabau, and perhaps most notably, the Stone Age Austronesians of the Mentawai Islands to the west. They include the proud hunter-gatherer Kubu of the Jambi province forests, leaderless 'children of the tiger', as well as the seafaring Malays of the south. In the south too stand strangely inscribed and decorated megalithic stones, testifying to another age. Here is the land of the notorious were-tigers of Kerinci, the stuff of nightmares, stalking their rain forest realm, as if real tigers were not enough to worry about.

◁
The banded leaf monkey (*Presbytis melalophos*) is one of Sumatra's most widely spread leaf monkeys or langurs. It comes in a variety of differently coloured subspecies, however. These monkeys move about in troops and have stomachs specially designed to digest the large quantity of foliage they consume. This one was photographed in West Sumatra.

Sumatra might well have been named 'The Island of Gold' for its gorgeous orchids, such as this giant tiger orchid spray (*Grammatophyllum speciosum*).

Yet more Sumatran 'gold'—the attractive fungus *Xeromphatira tenuipes*.

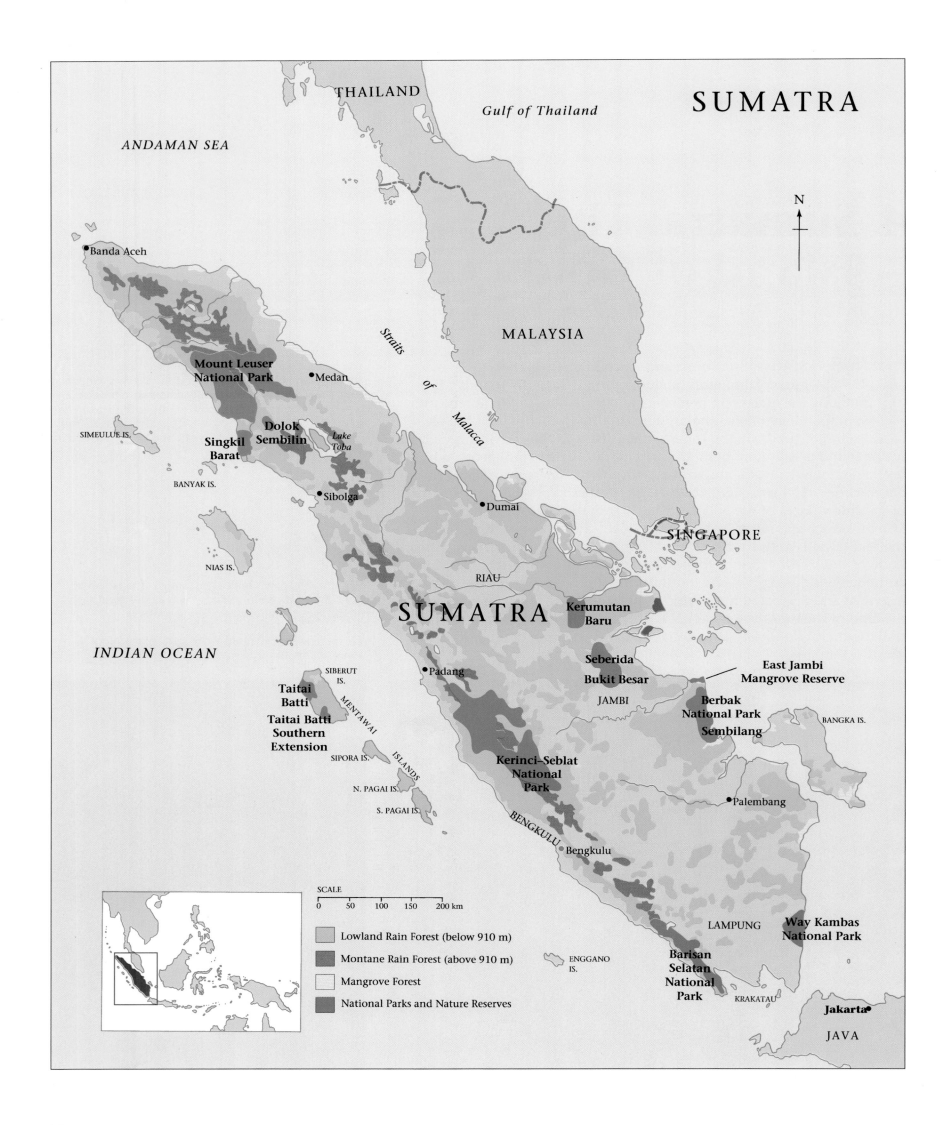

SUMATRA

THAILAND

Gulf of Thailand

ANDAMAN SEA

N

• Banda Aceh

MALAYSIA

Mount Leuser National Park

• Medan

Straits of Malacca

SIMEULUE IS.

Dolok Sembilin

Lake Toba

Singkil Barat

BANYAK IS.

• Sibolga

NIAS IS.

• Dumai

SINGAPORE

RIAU

SUMATRA

Kerumutan Baru

INDIAN OCEAN

SIBERUT IS.

Taitai Batti

Taitai Batti Southern Extension

MENTAWAI ISLANDS

SIPORA IS.

• Padang

Seberida

Bukit Besar

East Jambi Mangrove Reserve

JAMBI

Berbak National Park Sembilang

BANGKA IS.

N. PAGAI IS.

Kerinci–Seblat National Park

S. PAGAI IS.

• Palembang

BENGKULU

• Bengkulu

SCALE

0   50   100   150   200 km

☐ Lowland Rain Forest (below 910 m)

☐ Montane Rain Forest (above 910 m)

☐ Mangrove Forest

☐ National Parks and Nature Reserves

ENGGANO IS.

LAMPUNG

Way Kambas National Park

Barisan Selatan National Park

KRAKATAU

Jakarta•

JAVA

'Here be sweet-smelling woods of great utility, but also beastly cannibals' was what the early Venetian traveller Marco Polo reported in the thirteenth century from Sumatra, which he had mistaken for 'Lesser Java'.

On the one hand, travellers like Marco Polo praised the abundance of forests and 'costly products' such as gaharu and spices, as well as plentiful fish, sago flour, and potent palm-wine ('toddy'); on the other hand, they reviled the propensity of the native peoples for eating human flesh. However, it must be said here that this was a fairly infrequent and mostly ceremonial practice, being reserved more for symbolic victory sacrifices in wartime, or for punishment of wrongdoers.

Early morning at Mount Leuser National Park, one of Sumatra's major forest reserves.

A small palm (*Licuala* sp.) rises from the forest floor.

The giant chameleon (*Gonyocephalus dilophus*), photographed here at Mount Leuser, Sumatra, is actually an Agamid lizard. No true chameleons are found in South-East Asia.

Marco Polo probably did not fully comprehend the very civilized cultural impact of Sumatra on the surrounding region. The seafarers of Sumatra in particular have always exerted huge economic and cultural influence over the rest of Indonesia and indeed the region—both Hinduism and Islam entered Indonesia via the trading ports of Sumatra. The scattered islands of the Riau–Lingga archipelago long held political sway over Singapore Island and southern Malaya in pre-colonial times, as the Johor–Riau kingdom, while Palembang to the south was the seat of the great Buddhistic Srivijaya empire until about the eleventh century; the Malays, originally from Sumatra's Central Jambi area, the kingdom of Melayu, have spread through the region, notably in Malaysia.

The richness of the land has helped make the people of Sumatra particularly adept and knowledgeable in the use of medicinal plants. In Aceh to the far north, they use the bark of the tree which bears the yellow *campaca* flower (*Gardenia* sp.) for fever, while in North Sumatra, the bark of the fragrant-blossomed *kenanga* (*Cananga odorata*) tree is used for scabies, the leaves for skin-itch, and the seeds for stomach-ache. The local pharmacopoeia abounds with such examples.

The vegetation is as diverse as the people, ranging from lowland to alpine types (amid scenery featuring Indonesia's highest mountain outside Irian Jaya, and South-East Asia's second highest, Mount Kerinci, at 3800 metres, if you exclude Indo-China), and including some notable heath forest, swamp forest, and wetland areas, particularly at Berbak National Park on the east coast of Jambi province, close to the Straits of Malacca, Way Kambas National Park, and the Sembilang mangrove reserve in South Sumatra. There are even stands of pine (*Pinus* spp.) on the drier mountains in the north.

Up in the mountains hide several little-known and rare animals, such as the serow or mountain goat (*Capricornis sumatraensis*), the shaggy mountain giant rat (*Sundamys infraluteus*), or the striped Sumatran rabbit or hare (*Nesolagus netscheri*) confined to the Mount Kerinci area, the only rabbit or hare native to Indonesia. Among other mountain residents are the giant chameleon (*Gonyocephalus dilophus*), and the *bunga asakh* (*Mirabilis jalapa*) flower of Lampung province in southern Sumatra, always above 1400 metres. Thoughtfully, the *bunga asakh* opens its pretty little red, white, yellow, and violet flowers promptly at afternoon prayer time for the Muslims, reminding farmers to put down their tools and attend to their religious duties.

A Sumatran orang-utan (*Pongo pygmaeus*) and her baby revel in their forest world. This intelligent ape is threatened by habitat destruction and hunting.

There are disconnections within Sumatra. The northern and southern parts of the island, demarcated by the mighty eruption which created Lake Toba about 75,000 years ago, do not share the same animals. The orang-utan (*Pongo pygmaeus*), for instance, is to be found only to the north of the divide, while the Malayan tapir (*Tapirus indicus*) is common in the south. The mountainous north–south spine of the Barisan Range on the western side of the island also renders the low-lying swampy eastern scenery quite different from that found in the west. Islands like the Mentawai group on the west have been isolated for at least half a million years, allowing them to preserve or evolve quite different animal forms. Yet there are connections beyond the island, with Kalimantan and the rest of Borneo. Until about 7,000 years ago, Kalimantan and Sumatra were joined, which explains their similar ecologies.

In terms of species exclusive to itself, Sumatra is not as rich as Borneo, with only 12 per cent of its plant species, 10 per cent of its 196 mammals, and just over 3 per cent of its 602 birds endemic, but it does score well for diversity. It also boasts Indonesia's only remaining significant populations of wild tigers (*Panthera tigris*) and elephants (*Elephas maximus*), not to mention the world's tallest flower, the odiferous lily-related *Amorphophallus titanum*, and the world's largest flower, the *Rafflesia arnoldi*. Rajah Brooke's birdwing butterflies (*Trogonoptera brookiana*) are also found on the island. One of the great mysteries of Sumatra, however, is the absence of any leopards apart from the clouded type (*Neofelis nebulosa*).

Storm storks (*Ciconia stormi*) are only found in Sumatra and Kalimantan, having gone extinct in Peninsular Malaysia and southern Thailand. It occurs particularly in riverine and swamp forests.

Rajah Brooke's birdwing butterflies (*Trogonoptera brookiana*) are handsome natives of Sumatra, bejewelling the forest floor, especially along trails and by rivers, as they search for moisture. Their wing-span can reach 17 centimetres. They have been successfully bred in Sumatra.

The island is extremely rich in bird life, with about 450 resident species, the richest avifauna in the region after the island of New Guinea (including Irian Jaya). Storm storks (*Ciconia stormi*) are among the rarer species found on Sumatra. Ten species of hornbill, the great Argus pheasant (*Argusianus argus*), fairy bluebirds (*Irena puella*), trogons (*Harpactes* spp.), and barbets (*Megalaima* spp.) are among the delights of the Sumatran forest. Like all forest birds, many of these are extremely elusive and best known by their calls since they are hard to see: even some very experienced biologists have never yet caught sight of the great Argus pheasant despite hearing it all day long, while others need endless patience and a good ear to distinguish the call of the shy Malaysian rail babbler (*Eupetes macrocerus*) from that of the garnet pitta (*Pitta granatina*).

At first glance, hornbills seem to rank high on the male-chauvinism index for their unusual habit of sealing the female into her nest hole with mud and twigs, but the objective is really protection and nurturing, for the bird leaves a gap in the mud seal through which he lovingly feeds her until she is ready to emerge with her brood.

A rare shot of a group of white-winged wood ducks (*Cairina scutulata*). This endangered bird needs tall forest trees for nesting holes, although it feeds in swamps.

The great hornbill (*Buceros bicornis*) is one of Sumatra's 10 hornbill species. Its harsh 'bark' and guttural chucklings are a familiar sound in the forest.

The Malayan tapir (*Tapirus indicus*), common in southern Sumatra, prefers low-lying swampy areas. Because the animal feeds from low-level plants, grabbing them with its short 'trunk', it favours the kind of regenerating and secondary growth that is more typical of disturbed or logged forest.

A Sumatran speciality, but increasingly a rarity, is the white-winged wood duck (*Cairina scutulata*), recognized by its characteristic goose-like honking call. This attractive, black-and-white duck needs tall trees for nesting, generally in and around swamps, and it particularly favours the renghas tree whose poisonous sap is antipathetic to most human beings. Sumatra is this duck's only Indonesian location. There may be no more than 116 such ducks left on the island and perhaps no more than 211 in the world; they are known only from Assam in India, Bangladesh, Burma, Vietnam, and Thailand outside Indonesia, and even in these locations their highest known population is Assam's 15 pairs.

Sumatra's two major rain forest areas are the 7927-square-kilometre Mount Leuser National Park in the far north (in the Gayo language, Mount Leuser means 'the place where animals go to die', reflecting the belief among many traditional peoples that mammals climb mountains to die), and the huge 14 846-square-kilometre Kerinci–Seblat National Park spread across West Sumatra, Jambi, Bengkulu, and South Sumatra. Many

> The *tondi* [soul] becomes *begu* [ghost]
> The hair becomes *idjuk* [roofing material]
> The flesh becomes earth
> The bones become stones,
> The blood becomes water,
> The breath becomes wind.
>
> (An old Karo Batak proverb on the 'ashes to ashes' theme, linking human life and death with Nature)

The mist-shrouded highlands of the Kerinci–Seblat National Park. Just under half of Sumatra is still heavily forested.

visitors to Mount Leuser are, in fact, on their way to see the orang-utan rehabilitation centre at Bohorok, Bukit Lawang, or perhaps to white-water raft their way down the Alas River. The semi-tame orang-utans, many orphaned by hunters, are painstakingly taught how to be wild men of the forest once again, but experts disagree on the success, or even usefulness, of such ventures, other than the obvious 'PR' for forest habitat conservation.

Besides the marvellously shaggy red orang-utan, there are also more delicate primates to see, such as the white-handed gibbon (*Hylobates lar*), Thomas's leaf monkey (*Presbytis thomasi*), the banded leaf monkey (*Presbytis melalophos*), and the acrobatic siamang (*Hylobates syndactylus*), besides the somewhat less graceful macaques, both long-tailed (*Macaca fascicularis*) and pig-tailed (*Macaca nemestrina*). The bird-watching too is splendid, with about 500 species present.

The acrobatic siamang (*Hylobates syndactylus*) is a gibbon, although it is twice the size of most other gibbons. Like all gibbons, it moves through the treetops using its arms to swing—a process scientists call 'brachiation'. Siamangs form close-knit families, with the fathers playing an active role in baby care.

The pig-tailed macaque (*Macaca nemestrina*) is a familiar sight in Sumatra's kampongs, where it is often domesticated and trained to climb and pick coconuts for humans.

Kerinci–Seblat, stretching for 345 kilometres along the Barisan Range, offers the majestic Mount Kerinci (which last erupted in 1934), and Mount Tujuh with its pristine crater lake at 1996 metres. The larger mammals, such as tigers, clouded leopards, elephants, rhinoceroses, and sun bears (*Helarctos malayanus*) call the Kerinci forest 'home'. Here roams the largest group of endangered Sumatran rhinoceroses (*Dicerorhinus sumatrensis*) in the world (note, however, that it is not their habit to group in herds), about 400 of the world's total population of perhaps 800. Could there perhaps also be a lost tribe, the short hairy people known as *orang pendek* (the short people), whom the locals claim they have sighted in the forest? These possible hominids share a question-mark place in the Kerinci ecology with the *cigau* or lion-tiger and the *kuda liar* or wild horse also reported by local people, but not confirmed by scientists.

The Kerinci–Seblat National Park, its integrity bedevilled by intensive cinnamon farming, is also a case-study for the new approach to conservation which teams human needs with forest protection, which has now been adopted by the Indonesian government. To this end, the park is the site for an Integrated Conservation and Development Project partnering the government with the World Bank and the United Nations Development Programme. With Sumatra's population growing at an average of 3 per cent per annum, well above the annual average of 2 per cent, there is pressure on the natural environment and wildlife. Many observers feel that successful management of human conflict with Nature in Sumatra will signpost the way ahead for the rest of Indonesia.

Tales of Sumatra

Now listen, and hear that North Sumatra's Lake Toba, almost 1000 metres up in the highlands, represents the tears of a fish-princess at her human husband's broken promise.

When powerful fish-goddess Boru Saniang Naga married Toganaposo, a farmer and skilled martial arts exponent, she demanded only that he never again mention her fishy origins. Their son was called Samosir and grew up spoiled, lazy. Losing his temper one day, Toganaposo hit the boy, calling him the son of a fish. Regret came too late. A fiery volcanic eruption killed Toganaposo, and next, the heavens opened. It rained and rained and rained, until the 1145-square-kilometre, 450-metre-deep Lake Toba had been born, with an island in the middle of the lake—called Samosir.

* * *

Know that to build a swift *kora-kora* boat or *perahu* in Riau is a serious endeavour, requiring spirits of the forest to be placated before the desired tree is cut down. Once upon a time, the boats asked for human sacrifices before they would launch.

The story is told of the royal boat *Lancang Kuning* (Yellow Warship) built at Bengkalis Island by naval chief Umar. Umar's rival was the military commander Hassan, who not only vied with him for power but also for the woman Umar finally wed, the lovely Zubaidah. While Umar was away fighting the pirate enemies of the local raja, Hassan deceived the raja into delegating all authority to him. It had proved impossible to launch the newly built *Lancang Kuning* for some time already, despite the sacrifice of a live buffalo. Newly empowered by the raja, Hassan insisted on the sacrifice of the now pregnant Zubaidah herself, without consulting the ruler. The terrible deed was done, and sure enough, *Lancang Kuning* at once slipped obediently into the water, with poor Zubaidah bound to her keel.

Fragile Fantasy Flower

It appears at first in the form of a small round knob, which gradually increases in size. The flower-bud is invested by numerous membranaceous sheaths, which surround it in successive layers, and expand as the bud enlarges, until at length they form a cup round its base.... The inside of the cup is of an intense purple, and more or less densely yellow, with soft flexible spines of the same colour; towards the mouth it is marked with numerous depressed spots of the purest white.... The petals are of a brick-red with numerous pustular spots of a lighter colour. The whole substance of the flower is not less than half an inch thick, and of a firm fleshy consistence.

(Thomas Stamford Raffles, describing the *Rafflesia arnoldi*, known locally in Sumatra as the 'Devil's Betel Box', in 1818)

Nothing could illustrate the other-worldliness of the rain forest more dramatically than Indonesia's *Rafflesia* flower. Found in shady lowland rain forest, the *Rafflesia* is a celebrated botanical curiosity. Almost 1 metre wide and weighing about 9 kilograms, it is the largest bloom in the world, besides being one of the rarest and most endangered.

The repugnantly fleshy, still more offensively malodorous flower looks like a creation of science fiction, a moon-flower alien transplanted from a distant planet, a childish nightmare dreamed up by some Hollywood set-maker. But *Rafflesia* is real, particularly in Sumatra, although it is also found in Java, Borneo (Kalimantan), Bali, and Mindanao in the Philippines. At Batang Pulupuh, north of Bukittinggi in the highlands of West Sumatra, there is a special *Rafflesia* reserve where visitors may be fairly sure of seeing at least a *Rafflesia* bud. September and October are said to be the best months for a rare view of the full bloom. Another good location is the area around Taba Penanjung, near Bengkulu, off the main Bengkulu–Kepahiang road.

A *Rafflesia arnoldi* bud, at Mount Sago, West Sumatra., about one week before blooming.

The *Rafflesia arnoldi* flower unfurled, at the same site in Sumatra. The largest bloom in the world, it can reach 1 metre in diameter.

The *Rafflesia arnoldi* species, restricted to Borneo and Sumatra, is the most famous and the largest of 10–15 species (taxonomists are still arguing about this), all found only in South-East Asia. Four of these are found in Sumatra. The flower is an excellent example of how fragile some components of the rain forest ecosystem are, for its very survival is dependent on the survival of one particular vine, called *Tetrastigma*, related to the grape vine. If you did not know the very special role played by *Tetrastigma* in preserving *Rafflesia* for posterity, it would be the kind of unremarkable vine you might have no compunction in cutting down as you passed by. This is one of the factors that makes *Rafflesia* very vulnerable, and very rare.

The *Rafflesia* is a disembodied flower. A rootless, leafless, and stemless parasite, it drains nourishment and gains physical support from its host vine. Its only 'body' outside the flower consists of strands of fungus-like tissue growing inside the *Tetrastigma* vine. As Raffles himself so graphically described it, the *Rafflesia* first manifests itself as a tiny bud on *Tetrastigma* roots or stems, swelling over a period of 21 months to a cabbage-like head. Then, slyly, the head bursts around midnight, under the cover of a rainy night, to reveal a startling, lurid-red flower.

The flower's five leathery petals encircle a cauldron-like cup, which contains a spiked disc. Attached to the underside of this disc are either stamens or stigmas, depending on whether the plant is male or female. You could not miss a *Rafflesia* if you were near one: aptly tagged *bunga bankai* or 'corpse flower' by the Indonesians, it smells like rotting meat, which serves to attract carrion-scavenging flies and beetles. These pollinating agents will then crawl inside the flower and pollinate it, carrying pollen from a male to a female flower. The flower's glory is short-lived, lasting only up to a week, after which it withers and dies, collapsing on top of its own fruits.

Inside the leathery petals of the *Rafflesia arnoldi* is a cauldron-like cup containing a spiked disc which in a male carries stamens, in a female, stigmas.

It is one of Nature's miracles that such a rare flower should succeed in arranging to bloom synchronously with and close to another such flower of the opposite sex. Still more miraculous is the fact that *Rafflesia* seeds ever find their way to precisely the vine they need. Scientists are not sure exactly how these seeds are dispersed in the forest, or injected into their host vines. They may be carried on the feet of wild pigs, tapir, deer, squirrels, rodents, or jungle fowl, which like to feast on *Rafflesia* fruits, reputed to smell like rotten coconut.

It seems that the vine must be slightly torn or damaged in some way in order to receive the seeds into its interior, and it may be that the seed-carrying animals themselves do this damage. This would make the *Rafflesia* an even better case-study of rain forest 'networking', since the fates of hoofed animals like pigs or deer and of the vine *Tetrastigma* would all seem to be crucial to the survival of the flower. The *Rafflesia*'s need for large and medium-sized animals as dispersal agents, in turn, implies the need for protected reserves big enough to maintain such large animals.

It is hardly surprising that such an unusual bloom should acquire for itself a semi-magical reputation. Among the Sakai people of southern Sumatra, a *Rafflesia* infusion is said to be a powerful aphrodisiac for males, while other Indonesians use the bud to help women recover their youth, energy, and figure after childbirth. *Rafflesia* is also a choice ingredient in Indonesia's famous *jamu* range of herbal medicines and cosmetics.

This lesser tree shrew (*Tupaia minor*), seen here walking on a *Tetrastigma* vine of the type which *Rafflesia* parasitizes, may well be one of the agents which transfer *Rafflesia* seeds from the flower to a new host vine.

The Lord Whose Name May Not Be Spoken

The tiger (*Panthera tigris*) has disappeared from Bali, and from Java too, and may never have been in Kalimantan, but it still reigns supreme in the rain forests of Sumatra, where there are 466 of its kind.

Although on the small side by Siberian standards, the Sumatran tiger is a powerful predator held in terrified awe by Sumatran peoples. Some—notably the Kubu people of Jambi and the men of Kerinci—claim a special relationship with the tiger. The Kubu will never hunt a tiger as a result, while the men of Kerinci, in common with those of some other Indonesian communities, are renowned and feared for their reputed ability to transform themselves into were-tigers at will. In Jambi province, people bestow the honorific 'Datuk' on the great cat, as if to call it 'Sir Tiger'. As is true of many wild animals in traditional societies, it is considered bad luck to call the tiger by its real name, *harimau*, while in the forest, as this is likely to summon the animal from its hiding place. The roar of a tiger also signifies impending disaster in that area.

Riau province, although rapidly coming under human development programmes, is good tiger country, and Berbak National Park on the eastern coast of Jambi province fringing the Straits of Malacca is another location for tiger-spotting, as are the forests of Aceh province, in northern Sumatra. Other good tiger habitats are the Mount Leuser and Way Kambas national parks in the south.

The tiger typically stalks and hunts alone by night, travelling up to 30 kilometres. By day, it rests in the shade or in long grass close to a stream, enjoying an occasional splash in the water. It will eat a wide range of animals, both large and small, from fish and tortoises to its favourites, pigs and deer. Contrary to myth, it is not an infallible hunter, often failing to make a kill. It may target the young, the sick, or the old, to maximize its chances of success. Contrary to popular misconception, too, tigers rarely bite or claw their prey to death, unless the animal is small enough to dispatch with one bite through the neck. Although it is true that individual tigers differ widely in their killing styles, tigers generally hold their prey by the throat until it strangles, forcing it to the ground.

The Sumatran tiger (*Panthera tigris*) is a shy and secretive beast which hunts alone by night, ranging across as much as 50 square kilometres of territory. Sumatra is the last real refuge of the tiger in Indonesia. The animal is fully protected but hunting does go on, and Sumatran tiger-skins do find their way on to the market.

Characteristically, they will start eating the carcass from the hind quarters forwards.

By nature a silent, secretive, and cautious beast, the tiger is more anxious to avoid human beings than to confront them. Most naturalists feel privileged if they ever see a tiger in the wild—it is certainly one of the most thrilling sights the writer can recall. However, human incursions into tiger territory, a perceived threat to a female's cubs, or the onset of old age and consequently diminishing hunting skills, may induce a tiger to attack Man. Since the tiger usually hunts by night, and commonly attacks from the rear, night forest trekkers, in particular, are advised to keep their eyes and ears open.

There is much evidence to suggest that the tiger assesses its proposed prey's strength and size by his height as seen from the front. A fully erect human being facing the tiger is a discouraging prospect, but a crouching man or woman, a small child, or someone with his back turned, can seem invitingly small and manageable. This is why so many of the few recorded tiger attacks on Man turn out to have been on workers who are bending down, with their faces concealed—rubber tappers, for example—or who are squatting to defecate or urinate.

Sumatran tigers need very large areas to maintain even quite moderate populations, since a single male tiger usually controls about 50 square kilometres of territory and a female about 20 square kilometres. Biologists reckon that a minimum viable population of tigers, numbering about 60 animals, may need at least 750 square kilometres. Although solitary, tigers within a certain range—and a male's territory may overlap with a female's in many cases—keep in constant contact, mostly by scent. As can also be observed with pet cats, the tiger marks territory with a pungent scent produced in its anal glands and then mixed with the animal's urine as it sprays the area. These scent traces act as a kind of personal notice-board informing other tigers of their fellow tigers' whereabouts, direction

Tigers can swim and love to bathe in rivers. They are in fact quite intolerant of full sun, reflecting their prehistoric origins in northern Asia. Their diet is quite varied and includes fish, which they are adept at scooping from rivers.

of travel, state of health, sex, and state of sexual receptiveness. Experienced human hunters and trackers can also detect and read this scent when trailing tigers.

Roaring is another way tigers communicate with one another, although this does not happen very often. In the case of two males, particularly, this serves as a warning to keep out of each other's way, lest they be forced to confront each other and fight, again an extremely rare occurrence. If they do meet, they may well be able to recognize each other. Field researchers have found that individual tigers can be distinguished by the sharply defined differences in their face markings.

A tigress will normally bear a litter of 1–7 cubs, although probably only 2–3 will survive. She gives birth about every 2–3 years, after a 3–4 month pregnancy. Cubs are raised solely by the mother, without any contact with their father.

There is very strict protection for all tigers under the Convention on International Trade in Endangered Species of Wild Fauna and Flora (CITES), which Indonesia has signed, and also under an Indonesian governmental decree of 1972. None the less, because of their power, size (about 2.5 metres long and 185 kilograms in weight), dramatic markings, and sheer rarity, tigers are still considered the hunter's most desirable trophy. Their handsome skins also fetch a high price, usually several thousand US dollars, while their bones are sought after as important ingredients in a range of traditional medicines, particularly among the Chinese, who use powdered tiger bones or 'tiger bone wine' in the treatment of several ailments, including ulcers, rheumatic pains, typhoid fever, and malaria. Lamentably, supposed tiger-tooth pendants and the like are still commonly sold in souvenir shops throughout South-East Asia, but many of these actually come from dogs or leopards. With a mere 5,000 tigers surviving world-wide now, Sumatra's tigers have become more precious than ever.

These two Sumatran tiger cubs may have survived from a litter of seven. They are raised by the mother and may never know their father. The mother is fiercely protective and extremely dangerous to human beings while in charge of her cubs.

Pachyderm Power

The relatively small elephants (*Elephas maximus*) of Sumatra (about 3 metres tall and 5 tonnes in weight) were once a familiar sight at the opulent courts of Sumatran kings and sultans, who rode them to war and also set them against each other in gladiatorial contests for their own entertainment. So respected were these royal elephants that some mahouts or trainers were expected to die with them; one story from Aceh in northern Sumatra tells of a mahout being killed and thrown into the sea inside the dead animal's stomach.

Even today, some Sumatran villagers believe elephants cause lightning—by tossing it from their trunks, by sharpening their tusks, by shaking their ears, or through the trembling of their limbs. Many Sumatrans still dimly hark back to the echoes of ancient Hinduism, which revered Ganesha the elephant god, gentle lord of wisdom and culture, or to Buddhism's awe for the white elephant, an incarnation of Buddha himself.

It was under Dutch colonial rule that the Indonesian rulers first began to neglect the art of domesticating and training these massive beasts, Indonesia's largest land animals, as they became distracted by trading profits. A revival of that art now seems to be the best solution to increasing conflict of interest between humans and elephants in Sumatra. This conflict, despite complete legal protection for the Sumatran elephant since the 1930s, is one of the most spectacular examples of Man's unfortunate disharmony with Nature. As ecologists John and Kathy MacKinnon have put it, 'Many species go quietly to extinction, but not the elephant.'

Elephants need enormous space, preferring lowland forest. When they become overcrowded and their forest foods dwindle, and when their traditional seasonal migratory routes between lowland and mountainous

This is Indonesia's largest land animal, the Asian elephant (*Elephas maximus*), seen at Sumatra's Way Kambas National Park. The easiest way for the layman to tell the difference between an African and an Asian elephant is to know that the Asian species has much smaller ears. This is a male, but elephant society is largely matriarchal.

An elephant line-up—the Asian elephant moves around in herds of anything from 5 to 20 animals, usually all females and calves, led by an older female. Males only join them when it is time to mate.

areas are disturbed, these highly intelligent animals react and fight back, primarily by raiding human settlements and farms for cultivated crops such as rice, corn, and oil palm. They have been known to develop planned and reasoned strategies for dealing with the electric fences or other obstacles that human settlers have put in their way, particularly in southern Sumatra, which is home to more than 40 per cent of Sumatra's estimated elephant population of 2,800 animals. This total population is fragmented into 44 separate populations, nearly 30 per cent of which are less than 50-strong. Existing protected areas probably cannot accommodate more than about 2,500 elephants.

Needless to say, elephant herds pay little attention to the boundaries demarcating national parks and reserves from human habitat, so this is one case where protective measures for animals straying outside the national parks are needed. Education of the people and compensation for the losses they incur during elephant raids on their settlements are two such measures, but they are difficult to implement and perhaps not enough in themselves. Conversely, it could be said that humans have not paid enough attention to natural or ecological boundaries when demarcating national park boundaries.

Elephant social structure revolves around a group of 5–20 elephants, usually dominated by a female. While anthropomorphism should perhaps be avoided with wild animals, it is difficult to resist the impression of great tenderness in family relationships and overwhelming group loyalty within the herd. Particularly moving is the way one or two older females—'aunties'—will assist with the birth of a new calf, while the rest of the herd forms a protective circle around the labouring mother. The whole herd takes great care of the calf thereafter. Calves stay with their mother until about eight years old.

This baby elephant sheltering beneath its mother will probably be nursed for about two years. Most elephants give birth to one baby, only occasionally to two. The young elephant reaches sexual maturity at between 8 and 12 years old.

As with the rhinoceros and many other animals, careful logging need not threaten wild elephants, since logged forests often offer the animals even better feeding opportunities. An elephant needs about 135–300 kilograms of vegetation daily. Elephants particularly favour the leaves of plants like bananas and gingers, and young bamboo shoots, as well as fruit. There can be no mistaking an elephant feeding-ground—it resembles a battle-ground. Sumatran elephants also look for mud wallows, mineral earth-licks and, of course, water, about 225 litres of it a day. Unlike their African counterparts, they also need forest shade against the tropical midday sun.

That elephants can have their uses as tourist attractions and work animals is now being realized at the elephant training centres established since 1986—at Way Kambas National Park in Lampung province, and in the provinces of Aceh, Riau, and South Sumatra. By 1989, there were more than 50 elephants learning new skills from Thai experts at these centres, as were many prospective Indonesian mahouts. The idea is to use these elephants as a complement to, or supplement for, heavy machinery when transporting logs during selective logging operations, especially on terrain too tough for machinery to traverse. They need no spare parts and do far less damage to the vegetation surrounding desirable timber trees than do bulldozers and lorries. Unfortunately, the needs of industry may be much less than the supply of wild elephants.

The domesticated animals can also be used to control and tame troublesome wild elephants. There is nothing like an elephant football match for getting the educational message across to the Indonesian public that elephants are not necessarily a crop-devastating pest to be eliminated. Trained elephants are also useful as tourist–safari transport, for park boundary inspection and park patrols, and for the transportation of construction materials to remote guard-posts within parks.

Indonesia demonstrated its commitment to preserve the Sumatran elephant with a massive elephant 'drive' at the end of 1982, which pushed the animals out of the way of newly arrived transmigrants in South Sumatra and so avoided conflict. In a huge and expensive military-style operation deploying soldiers, helicopters, and speedboats, 232 of the pachyderms were driven into the 750-square-kilometre Air Sugihan sanctuary. But such drastic action does not always work, for elephants frequently try to make their way back to their old territory. The Sumatrans continue to seek some sort of accord with these splendid giants of the forest.

Training for a job in the modern world—Asian elephants at Way Kambas National Park, learning to perform tasks such as carrying visitors and goods. As Man intrudes more and more into the elephant's habitat, this may be the only way to avert the animal's ultimate destruction.

Gentle Fossil

It is surprising how much the world knows about the two African species of rhinoceros—the white (*Ceratotherium simum*) and the black (*Diceros bicornis*)—but how little about the three Asian species—the Indian (*Rhinoceros unicornis*), the Javan (*Rhinoceros sondaicus*), and the Sumatran (*Dicerorhinus sumatrensis*) rhinoceroses—which are just as rare, if not more so. One is forced to conclude that this is simply because the African rhinoceros is easier to see and photograph, out on open plains, than its Asian counterparts, which are much less aggressive, far more shy, and lurk in dense, often mountainous forests, virtually invisible most of the time.

The Sumatran rhinoceros has been declared one of the world's 12 most endangered species by the International Union for the Conservation of Nature and Natural Resources (IUCN), and it is officially protected in Indonesia. In the wild, there are perhaps only about 800 Sumatran rhinoceros left world-wide, certainly no more than 1,000, and about 700 of these are on Sumatra itself, with possibly a few remaining in Kalimantan. Sumatra's Kerinci–Seblat and Mount Leuser national parks are key rhinoceros locations.

Sporting a scanty coat of russet-coloured hair, the two-horned Sumatran species shares its ancestry with the prehistoric woolly rhinoceros (*Coelodonta*), tracing its descent to an early tapir-like creature living about 40 million years ago. Only a hundred years ago, this fossil roamed as far abroad as the Himalayan foothills of India and across South-East Asia, but now it is confined to Indonesia and a few small pockets in Malaysia.

The Sumatran is a mini rhinoceros, the smallest of the world's five rhinoceros species; it is about 100–135 centimetres tall, 250–280 centimetres long, and weighs around 800–1000 kilograms. It does not have the strongly folded, armoured appearance of the single-horned Javan rhinoceros, which is much bigger and about twice as heavy.

They are scarcely smaller than elephants. They have the hair of a buffalo and feet like an elephant's. They have a single large, black horn in the middle of the forehead. They do not attack with their horn, but only with their tongue and their knees; for their tongues are furnished with long sharp spines, so that when they want to do any harm to anyone they first crush him by kneeling upon him and then lacerate him with their tongues. They have a head like a wild boar's and always carry it stooped towards the ground. They spend their time by preference wallowing in mud and slime. They are very ugly brutes to look at. They are not at all such as we describe them when we relate that they let themselves be captured by virgins, but clean contrary to our notions.

(Marco Polo's view of the Sumatran rhinoceros, from R. E. Latham's translation of *The Travels of Marco Polo*, 1958)

Sumatran rhinoceroses (*Dicerorhinus sumatrensis*) now number less than 1,000 world-wide, with the largest population on Sumatra itself, notably in Mount Leuser National Park. This one was photographed at Torgamba in Riau province.

Like all rhinoceroses, the Sumatran rhinoceros needs to wallow regularly in mud-baths. Not only does this keep the animal cool, but it also helps it to repel and to remove flies and parasites, which it scrapes off by rubbing itself clean on tree bark.

It is sometimes difficult to make out the Sumatran species' distinguishing second horn, as this is often worn down to a mere bump by rubbing against tree-trunks and vegetation. The animal is a gentle herbivore and exhibits a certain charm: the young are known to bleat and squeak much like lambs or piglets. It browses on the leaves and twigs of more than 100 plant species, and it enjoys eating fruit, especially mangoes and mangosteens. Very possibly, the rhinoceros plays an important part in distributing the seeds of such fruit, via its droppings.

Some researchers have observed the rhinoceros eating so much of a particular mangrove tree's tannin-rich bark (*Ceriops tagal*) that its urine has been stained orange. As tannin is used in some drugs to treat human diarrhoea, it could be that the rhinoceros also uses tannin to treat intestinal ailments.

Like the tiger, the Sumatran rhinoceros requires very large areas of land—100 square kilometres per rhinoceros—to survive. It seems this need not all be virgin forest, however, for there is some evidence that certain plant growth in secondary regrowth or semi-cleared areas provides good fodder for the rhinoceros.

Largely a solitary wanderer, except when mating—a mother and calf will also stay together for about 18 months—the Sumatran rhinoceros is extremely shy and therefore very difficult to observe in the field: at least one well-respected researcher chose to base his thesis on the species almost entirely on plaster casts of the beast's footprints, with only a single sighting of the actual animal. Rhinoceroses have no difficulty communicating with one another, however: they 'speak' through smell, leaving urine squirts and communal dung piles as notice-boards with messages for other rhinoceros. As with the tiger, the difference among scents may indicate to a male whether or not a female is on heat or

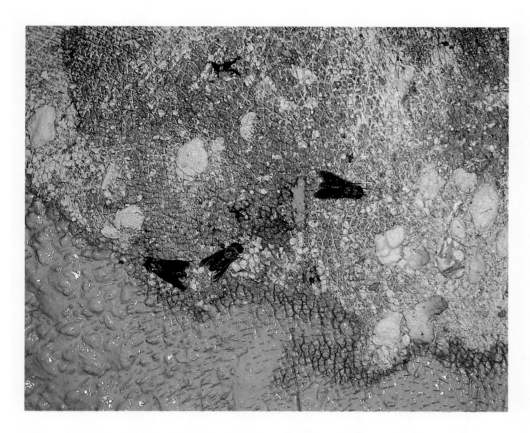

This close-up shows the rhinoceros's muddied hide, with irritant flies trying to penetrate the mud coat.

pregnant. The animals also socialize at communal salt-licks and mineral springs, where they seek minerals to remedy deficiencies in their normal diet.

The Sumatran rhinoceros is hunted illegally for its horn and other body parts prized in Asian folk medicine, chiefly as a fever tonic (not often as an aphrodisiac, as is popularly believed). The animal's often fixed patterns of movement along known trails make it highly vulnerable to pit-trap poaching; the hunter has only to wait for the animal to fall in and cruelly impale itself on a spear. Yet the animal's so-called horn is nothing more magical than an agglomeration of keratin, the very material that makes up our own hair and finger-nails, quite unlike cattle horn or deer antler.

Just as threatening to the rhinoceros's survival has been the steady clearance of its natural habitat—dense forest—for agriculture, industry, and human settlement. The pressure has driven most rhinoceroses uphill to rugged mountainous areas, despite their physiological need for swampy mud-wallows to protectively coat their hides against parasites and dry-cracking, and to cool themselves. Many individual rhinoceroses are now too isolated from each other to form viable breeding populations.

In 1985, amid controversy over the wisdom of captive-breeding, operations were launched to save the species by breeding it in zoos to maintain its gene-stock. Very few Sumatran rhinoceroses had been held in captivity—the last one died at Copenhagen Zoo in 1972—and little was known about the animal's reproductive cycle. Since 1985 there have been joint captive-breeding ventures between the Indonesian government and the Howletts and Port Lympne Zoo Parks group of England, and the American Association of Zoological Parks and Aquaria of the USA.

As a result of these efforts, about 11 rhinoceroses have been taken out of Sumatra and distributed amongst Indonesian and foreign zoos. Breeding pairs are now established in captivity and the long wait for offspring began in the late 1980s. Pregnancies among Sumatran rhinoceroses are thought to occur only once every four years; they last 14–19 months and result in only one calf. Because none of these captive-breeding attempts has yet borne fruit, the preferred approach to saving Sumatran rhinoceros threatened by development now is to translocate them to safer forested areas. Meanwhile, the battle to protect the rhinoceros in the wild continues.

Land of the Monkey Spirit

It is a long, long way from the satellite dishes of Jakarta city rooftops to the Mentawai Islands off the western coast of Sumatra: tens of thousands of years away, in terms of human life, but much longer in terms of wildlife.

Isolated from the mainland for perhaps half a million years, Siberut ('the mouse') Island has slept amid the Mentawai archipelago, locked in a time-warp. As a result, more than 60 per cent of the animals and 15 per cent of the plants on Siberut are endemic. These rare animals include monkeys, tree squirrels, flying squirrels, civets (Viverridae), frogs, and reptiles, and a tiny scops owl (*Otus* sp.). Their human companions, chiefly the Sakkudai, are equally at ease in the rain forests of Siberut. They live a cashless, classless, pre-Hindu Stone Age lifestyle, collecting sago, taro tubers, and plantains, hunting animals with bow and arrow, collecting forest produce, and rearing pigs. Their simple needs make minimal impact on the forest, which they respect and conserve for future use.

The people of Siberut have developed a relatively sophisticated and substantial oral pharmacopoeia of medicinal plants, their efficacy categorized according to their smell, taste, and colour; a fishy smell, for example, is associated with coolness and therefore good health, while several sour plants are considered good cures for influenza.

Of special interest are Siberut's four endemic species of primate: the *bilou* or Mentawai gibbon, also known as the Kloss gibbon (*Hylobates klossi*), a primitive black gibbon whose morning call is particularly soulful; the long-tailed, white-faced *joja* or Mentawai langur or leaf monkey (*Presbytis potenziani*); the *simakobu* or pig-tailed langur (*Simias concolor*), which comes in either a slate-grey or a golden form; and the *bokoi* or Mentawai macaque (*Macaca pagensis*), a macaque similar to the mainland's pig-tailed macaque (*Macaca nemestrina*). The *joja* and the *simakobu* are unusual among 'Old World' monkeys in forming human-like pair bonds with their sexual partners; in the *joja*'s case, the relationship is permanently monogamous, whereas the *simakobu* may also form harems.

No other island the size of Siberut—4480 square kilometres—can boast as many as four endemic primates. They are important in human life, both as food and for ritual or ceremonial purposes. The people hunt them, and monkey skulls adorn the native villages or traditional longhouses or *uma* so that the monkey souls may call out for more monkey company, luring their relatives to join them. The ornately tattooed people of Siberut do not hunt monkeys lightly, however. In their animist world, the whole of Nature—every plant, animal, rock, or stone—has a soul. Before they dare venture forth with their bows and poisoned arrows, their *kerei* or medicine man must perform various rituals to placate the monkeys or other prey they intend to kill. Gibbons are carefully protected and are only hunted under certain rules, while the golden form of the *simakobu* is never hunted, for it is believed to be a ghost. In this way, humans and monkeys attain balance together.

For the Mentawaian people, everything revolves around the soul and the feeling of 'rightness' in any action or any object. They believe the soul likes to take things slowly and pleasantly, without undue aggression; thus, their children are rarely scolded or disciplined. Everything they make is painstakingly crafted and ornamented, for they also believe that a magical 'rightness' derives from elegant handiwork.

The Mentawai or Kloss gibbon (*Hylobates klossi*) is a primitive gibbon with a soulful morning call. It is endemic to Siberut Island. Traditional controls on the hunting of gibbons have conserved it effectively until recent times.

In 1981, the whole of Siberut Island was declared a 'Man and Biosphere' reserve by the United Nations Educational, Scientific, and Cultural Organization (Unesco). Aware of the cultural and biological treasure that Siberut boasts, and of its fragility as mainland culture intrudes more and more on the island's people, the government has set aside 12.6 per cent of the island as a protected area, in two separate moves, in 1976 and 1981. Almost one-third of this area lies within the Taitai Batti wildlife reserve. In 1991, the government declared its intention to extend this so that 36.6 per cent of the island would be protected, and in 1993, it further banned all logging on the Mentawai Islands, which include Siberut.

5 Java: Among the Volcanoes

WHERE does the soul of Java lie? In the mystic *kraton* or palace cultures of Central Java, amid the timeless clangour of the gamelan orchestra; in the pre-Islamic and even pre-Hindu societies clinging to the slopes of ancient volcanoes; among the intricately organized fretwork of lush green rice terraces in the countryside; in the teeming lowland forests of Ujung Kulon to the far south-west; among the misty tree-ferns of highland Cibodas in the central west; or in the traffic jams of high-rise Jakarta? Probably in a complex combination of all these. As for the physical landscape that houses this soul, it is above all a monument to centuries of human endeavour, a craftsman's reworking of Nature's original handicraft.

◁
Paradise made real at Telaga Warna ('Pool of Colours'), a volcanic lake in Java surrounded by montane forest.

The extinct volcanic cone of Mount Pangrango soars 3019 metres above the lush forests of the Mount Gede–Pangrango National Park in Java.

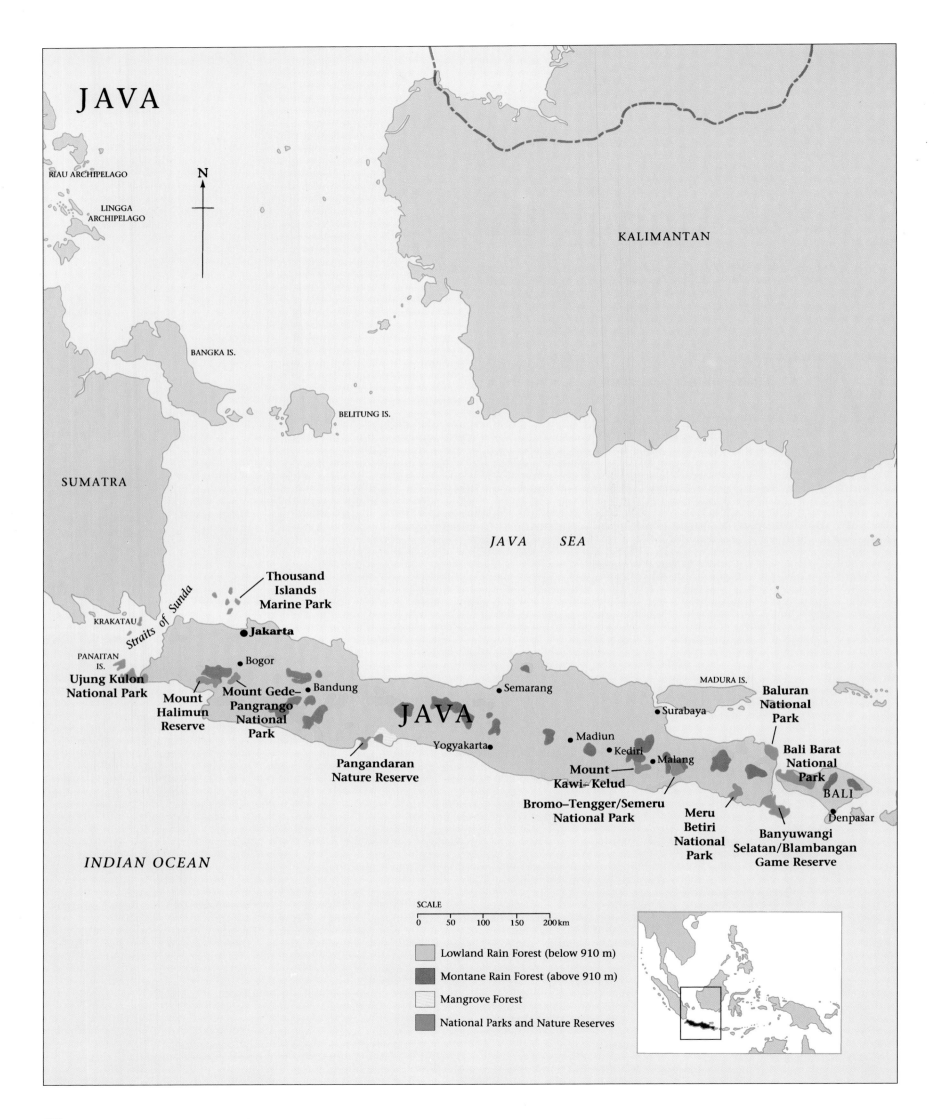

JAVA

RIAU ARCHIPELAGO

LINGGA
ARCHIPELAGO

N

KALIMANTAN

BANGKA IS.

BELITUNG IS.

SUMATRA

JAVA SEA

Thousand
Islands
Marine Park

KRAKATAU

Straits of Sunda

Jakarta

PANAITAN
IS.

Bogor

MADURA IS.

Ujung Kulon
National Park

Mount
Halimun
Reserve

Mount Gede–
Pangrango
National
Park

Bandung

Semarang

Baluran
National
Park

JAVA

Surabaya

Madiun

Pangandaran
Nature Reserve

Yogyakarta

Kediri

Malang

Bali Barat
National
Park

Mount
Kawi–Kelud

BALI

Bromo–Tengger/Semeru
National Park

Meru
Betiri
National
Park

Denpasar

Banyuwangi
Selatan/Blambangan
Game Reserve

INDIAN OCEAN

SCALE

0   50   100   150   200km

Lowland Rain Forest (below 910 m)

Montane Rain Forest (above 910 m)

Mangrove Forest

National Parks and Nature Reserves

Who are, or were, the people who have wrought this marvel? The murky origins of Java Man about a million years ago are best left for the palaeontologists to decipher. We are on firmer ground in saying that the modern Javanese come from Mongoloid stock speaking Austronesian languages, and moved into Java from the north about 4,500 years ago. The great *Ramayana* epic of Hindu India, possibly dating back to the third century BC, mentions Java as 'Yava-dvipa' (Barley Island), a land rich in grains and gold. Evidence from the early centuries AD shows that many travellers from China as well as India were already trading with Javan ports for various goods: spices, tortoiseshell, rhinoceroses, and elephants. The Hindu god Siva and Lord Buddha had been merged into a single harmonious faith, but by AD 780–850, the inspiration for the architects of the magnificent monument of Borobudur in Central Java had become wholly Buddhist. Somewhere around AD 900, the focus of cultural and political power moved from Central to East Java. By the fourteenth century, the mightiest kingdom Indonesia would ever see was established in Java—the short-lived Majapahit.

By 1478, Islam was on the march, and had converted Javanese rulers by the sixteenth century, leaving only pockets of the old religions on Java— Hindu enclaves like the Tenggerese highlanders of East Java and most of the Balinese, still surviving today. Even older survivors are the pre-Hindu cultures of the Badui people of West Java (it is said they are clairvoyant and can fly), and the Bali Aga of Bali. The biggest change of all, however, was heralded by the arrival of the Dutch colonists in the early seventeenth century, who imposed their exploitative commodity and plantation-agriculture based economy on the local people.

The rusa deer (*Cervus timorensis*) are common on Java, but being shy and nocturnal, they are not easy to see.

The muntjac or barking deer (*Muntiacus muntjak*) is a 'missing link' between the mouse deer and true deer. Like the mouse deer, the muntjac has long canine teeth and short pointed antlers.

Such changes had an impact on the natural landscape. Against the backdrop of a densely distributed and fast-growing human population, Java's forests are doubly precious for their scarcity. Today, only 7 per cent of the island is forested. Rain forest has retreated into pockets mainly in the extreme west and south of the island. Despite this, much remains to entrance the naturalist. As the most significant haven for a viable population of the very rare Javan rhinoceros (*Rhinoceros sondaicus*), the island merits attention.

In terms of its wildlife, Java ranks respectably, with 19 endemic mammals and 29 endemic birds, including the attractive little Java sparrow (*Padda oryzivora*), so popular as a cage-bird that it is now a familiar sight all over the world. The leopard (*Panthera pardus*), long gone in Sumatra, still pads Java's forests, particularly at Meru Betiri in the east, and there are more than 498 species of bird present on the island. Deer are plentiful, notably the sambar (*Cervus unicolor*) and the barking deer or muntjac (*Muntiacus muntjak*). Among Javan deer, the Bawean deer (*Axis kuhli*) might be described as a 'super-endemic', since its tiny remaining wild population of 300 is found only among the teak plantations on the small island of Bawean, north of Java, making it one of the world's rarest deer.

Fingered by some as possible competitors with the Javan rhinoceros for space and fodder, Java's wild banteng cattle (*Bos javanicus*) are nevertheless a handsome sight. A fully grown banteng may stand as tall as 1.7 metres and weigh around 800 kilograms. Bulls are dark brown to black, while the cows are a deep honey colour, but both are white-socked when adult. Not strictly deep forest animals, banteng often graze on clearings at the edge of the forest. Here, the risk of interbreeding with domestic cattle threatens their genetic survival.

In remote and wooded wilderness regions, the eerie, repetitive, and anxious trill of the endemic Javan barred owlet (*Glaucidium castanopterum*) may be heard with a bit of luck. By contrast, the rare Javan scops owl (*Otus angelinae*) of the West Java hill forests is more likely to be seen than heard, being a rare 'silent' owl.

Usually, a single male banteng (*Bos javanicus*) leads a herd of 25, including young males, cows, and calves. Banteng are largely nocturnal.

Java's rain forests yield to the bird-watcher some of the woodpeckers, pittas, barbets, bulbuls, and babblers common to similar forests in Sumatra, Borneo, and Malaya, but also three endemic barbets, seven endemic babblers, and three magnificent non-endemic hornbills—the rhinoceros (*Buceros rhinoceros*), the wreathed (*Rhyticeros undulatus*), and the pied (*Anthracoceros albirostris*). So industrious has the Javanese farmer been over the past 2,000 years that the best forest is now restricted to high altitudes not amenable to agriculture, for 63 per cent of the land has been put under cultivation.

Dominated by at least 17 still-grumbling volcanoes, the highest being the towering 3676-metre Mount Semeru, the plains of Java present a lattice-work of rice terraces and irrigation systems. The volcanoes slumber only temporarily: they can erupt any time, as did Mount Galunggung in West Java in 1982, after centuries of silence. Yet just under half the people of Java are farmers, and they continue to live close to the slopes of these fiery cones, trading their security for the fecund soils formed by the volcanic ash. In some areas, mostly in East and Central Java, only the hardy conifer-like cemara tree (*Casuarina junghuhniana*), a casuarina, can flourish in the ash, usually at heights above 1400 metres.

Proceeding north and east across Java, the land becomes harsher and drier and gives way to barren limestone slopes, dry deciduous forests, and open savannah grasslands. Yet much of Java, especially the western and central regions, is as steamy and wet as sensationalist colonial textbooks once stereotyped the tropical 'jungle', with hilly areas receiving 4000–7000 millimetres of rain a year. Towns like Bogor in West Java, with its gorgeous botanic gardens, put up with a world record of more than 320 thundery days in a single year. Bogor, or Buitenzorg as the Dutch knew it, at the foot of the Puncak mountain pass, was the place whose beauty the famous British colonial administrator, Sir Stamford Raffles, found so enthralling that he declared he and his beloved wife Olivia had spent the happiest days of their lives there. Raffles' aide-de-camp, Captain Thomas Travers, wrote of the area: 'There is certainly nothing to compare to it on

The male wreathed hornbill (*Rhyticeros undulatus*) has a call like a double bark. Compared with that of other hornbills, its casque is fairly small. This is one of only three hornbill species found on Java.

The banded pitta (*Pitta guajana*) is distinctive for its gorgeous colouring. This tiny and shy bird can be found up to altitudes of around 900 metres or so.

Tree-ferns like this one (*Cyathea contaminans*) at Cibodas flourish in the highlands, where they can grow to great heights. This is one of the tree species that give montane forest its characteristic appearance.

Most of Java's many volcanoes are only temporarily asleep. Here the Semeru volcano shows its power with a new eruption.

Plants of the rare *Balanophora* genus like this one, photographed at Cibodas, can only live by parasitizing the roots of trees.

▷
Parts of the Mount Gede–Pangrango National Park at Cibodas feature beautiful waterfalls, especially at Cibeureum—a wonderful habitat for ferns, mosses, and pitcher plants.

this island. The scenery is beautiful, between two rivers which have a rapid and meandering course over beds of stones in valleys abounding with luxurious verdure, and in sight of lofty mountains which may be said to lose themselves in the clouds.'

Up in the misty Puncak are the dripping tree-ferns of the montane botanic gardens at Cibodas, gateway to the spooky beauty of Mount Gede. Interestingly, montane forest areas such as these are important, not only for the species growing there, but also as seed reservoirs for many other species living at lower altitudes.

Java's most important rain forest areas are to be found in its national parks and nature reserves. The 786-square-kilometre Ujung Kulon National Park, first declared as a smaller protected area in 1921, is located at the south-western tip of the island, beyond which lies the infamous Krakatau volcano, now a reserve in its own right. The 580-square-kilometre Meru Betiri National Park, which was until very recently the stalking ground of the now extinct Javan tiger, is in the remote south-east of the island, while the 150-square-kilometre twin volcano park of Gede–Pangrango is in the west. The 400-square-kilometre Mount Halimun Reserve, adjacent to Gede–Pangrango, is the preserve of the spirits according to the folklore of the Badui people, who hold the reserve in great awe, respecting its integrity. Halimun is also a particularly good place to view rare and delicate primates such as the endangered silvery Javan gibbon (*Hylobates moloch*), also known as the grey gibbon, endemic to West Java (just 1,000-strong now), the Javan leaf monkey or *surili* (*Presbytis comata*), also known as the Javan grey langur, and the Javan silvered leaf monkey or *lutung* (*Presbytis cristata*), distinctive for its shaggy *coiffure*.

Javanese 'edelweiss' (*Anaphalis javanica*) is a large daisy-like flower found high up on mountains. The silver-leaved flowers are seen here on the grassy slopes of the Gede–Pangrango area.

Gede–Pangrango is known for its open fields starred with clusters of Javanese 'edelweiss' (*Anaphalis javanica*), also known in Sumatra and Sulawesi, and on Bali, in fact an outsize daisy, as well as other alpine and temperate-zone plants. The area is a treasure house of 208 orchids (of Java's total orchid repertoire of 731 species), 4 of which are endemic to Mount Gede. Various species of the quaint, carnivorous pitcher plant (*Nepenthes* spp.) are also on show. Trees vary from oaks (*Quercus* spp. and *Lithocarpus* spp.), chestnuts (*Castanopsis* spp.), and laurels (Lauraceae) to the white-flowered puspa tree with its red flushes of new leaf growth.

The Javan silvered leaf monkey or *lutung* (*Presbytis cristata*) can be spotted easily, thanks to its shaggy hairstyle. It is found only on Java and Bali.

Close-up of Javanese 'edelweiss'.

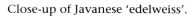

Begonias and an endemic pink-flowered impatiens (*Impatiens radicans*)—a balsam or touch-me-not, also known as 'Busy Lizzie'—are abundant at ground level in these mountain fields. Higher altitudes are given over to moss forests and landscapes of gnarled dwarf trees, swirling with mist. The red-and-black warty toad (*Cacophryne cruentata*) is known only from the Gede area, as is the volcano mouse (*Mus vulcani*), apart from its Sumatran location at Kerinci. Snuffling around the forest floor is the aptly named Javan stink badger or teledu (*Mydaus javanensis*).

The Javan kingfisher (*Halcyon cyanoventris*) is endemic to Java and Bali. Its Indonesian name *cekakak* mimics the sound of its loud chattering cry. It does not always feed on fish, being equally happy with insects and lizards, but it is seen here showing off its diving prowess.

Birds abound: the park is a safe retreat for the endangered Javan hawk-eagle (*Spizaetus bartelsi*) with its long tail and raised crest, for peregrine falcons (*Falco peregrinus*), and Himalayan swiftlets (*Collocalia brevirostris*) clinging to the walls of Mount Gede, and for the endemic red-tailed fantail (*Rhipidura phoenicura*), as well as 11 other endemic birds, such as the Javan chestnut-bellied partridge (*Arborophila javanica*), the large purple-and-turquoise Javan kingfisher (*Halcyon cyanoventris*), also found on Bali, the red-fronted laughing thrush (*Garrulax rufifrons*), the brown-cheeked Javan fulvetta (*Alcippe pyrrhoptera*), the brilliant green, yellow-and-red white-flanked or Kuhl's sunbird (*Aethopyga eximia*), the pygmy tit (*Psaltria exilis*), and the brown-throated (*Megalaima corvina*), blue-crowned (*Megalaima armillaris*), and black-banded or Javan (*Megalaima javensis*) barbets.

At Java's furthest south-western tip lies the important Ujung Kulon National Park, home to the Javan rhinoceros (*Rhinoceros sondaicus*).

The Cigenter River in Ujung Kulon. Crocodiles, a rarity in Java, can be found here.

The flying frog (*Rhacophorus* sp.) uses the webbing on its feet as a gliding membrane.

Ujung Kulon clings to the mainland by the thread of a narrow isthmus. Although its mud slides, tracks, and dung are fairly evident in the park, it is not easy to sight a Javan rhinoceros, so it is better here to concentrate on the equally rare black giant squirrel (*Ratufa bicolor*), the delicate mouse deer (*Tragulus* spp.), flying lemurs (*Cynocephalus variegatus*), black panthers (*Panthera pardus*), which are actually just black leopards, binturongs (*Arctictis binturong*), which are largely arboreal 'bear-cats', fishing cats (*Prionailurus viverrinus*), or the 1-metre-long monitor lizards (*Varanus salvator*), mini-dinosaurs, gliding flying frogs (*Rhacophorus* spp.) kicking strongly with their back legs to alter their direction as they fly from tree to tree, the crab-eating macaque monkeys, and hornbills. There are also crocodiles (*Crocodylus porosus*) in the rivers of Ujung Kulon, a rarity in Java. It is an interesting indication of the interrelatedness of things in the forest that there were tigers here until as late as the 1970s, when disease hit the deer population which would have been prime tiger prey.

Other parks are suffering from varying degrees of human population pressure on their resources. Future solutions can be glimpsed in projects such as the ongoing trial of nursery propagation for wild plants used for traditional medicines at Meru Betiri, a partnership between the authorities and the local people. That Nature is supremely resilient even in the face of great adversity, and that hope should never be abandoned, is amply demonstrated in the case of Krakatau. By extension, perhaps this may be taken as fuel for optimism about populous Java's endangered rain forests. Since its obliteration and violent reconstruction during the massive volcanic eruption of 1883, Krakatau has undergone a miraculous resurrection.

Wildlife and plants have recolonized the area since its virtual sterilization by the eruption. Only 14 years after the cataclysm, more than 132 species of insect and bird and 61 species of plant were recorded at Krakatau. Within 100 years after the eruption, the tally had grown to a total of over 1,200 life-forms, including more than 250 plants. Joint Australian–Indonesian scientific expeditions in 1984, 1985, and 1986 recorded the addition of another 15 new species of five different animal groups since studies conducted in 1933. The initial impetus came from sea- and wind-borne seeds or animals—a spider was carried in by the wind just nine months after the eruption—and their impact, in turn, was multiplied by the energies of fruit-eating bats and birds, recycling life through their guts. Life will continue to flourish in the Krakatau area. Secondary forest is already visible, but it will take many human lifetimes before true rain forest can re-establish itself there.

The Armoured One

The Javan rhinoceros (*Rhinoceros sondaicus*), or lesser one-horned Asiatic rhinoceros, is probably the rarest large mammal on Earth today, with none found in any of the world's zoos.

Records of the Javan rhinoceros date back to T'ang Dynasty times in China (AD 618–906). Documents of that time mention the export of rhinoceros horn from Java. A relief sculpture at the Cambodian temple of Angkor Wat, dating from about the twelfth century AD, also clearly depicts a single-horned Javan rhinoceros. In the first half of the nineteenth century, the Javan rhinoceros was found in Malaya, Burma, Thailand, Indo-China, possibly south-western China, and parts of northern India, as well as in Java and Sumatra. Some remnant animals apparently still linger in southern Vietnam, but otherwise it seems the Javan rhinoceros no longer exists across this range.

Apart from perhaps seven Javan rhinoceroses in Vietnam, only one other viable population of approximately 60 animals now remains—in Java's Ujung Kulon National Park. Animals concentrated in a single location in this way are extremely vulnerable to natural disasters (such as another catastrophic eruption of the Krakatau volcano near the park), drought, poaching, demographic instability, disease, and inbreeding. This vulnerability was highlighted dramatically in 1982 by the mysterious death of five Javan rhinoceroses in Ujung Kulon, possibly due to a cattle-borne, anthrax-like infection.

The number of rhinoceroses at Ujung Kulon doubled from the late 1960s, largely due to Professor Rudy Schenkel's successful joint effort with the Indonesian authorities to improve management and quash the rampant poaching. By 1980, the Javan rhinoceros population estimate of around 28 in 1967 had risen to over 54. But it is possible that a fixed area like Ujung Kulon may also have a fixed, and limited, maximum carrying

The Javan rhinoceros (*Rhinoceros sondaicus*) is among the world's rarest animals, with only 60 surviving, in Java's Ujong Kulon National Park.

The one-horned Javan rhinoceros is larger and much less hairy than the Sumatran rhinoceros, having more in common with the Indian species. The folded, armoured appearance of its hide is characteristic.

capacity for the Javan rhinoceros, so that the population can be expected to level out from now on.

During the period 1747–9, Javan rhinoceroses in Java were so numerous and the damage they caused to agricultural plantations was so heavy that the government of the day paid a premium of 10 crowns for every ánimal that was killed. This offer was not withdrawn until two years later and after 500 rhinoceroses had been bagged. The animal was still common in West and Central Java in the nineteenth century, especially in high or sparsely populated regions, but no more. Official protection for the Javan rhinoceros has been extended since 1908, but at first there was little supervision or implementation of the law on the ground, and poaching continued, with 42 animals taken between 1929 and 1967 alone. Since 1967, the World Wide Fund for Nature (WWF) and the International Union for the Conservation of Nature and Natural Resources (IUCN) have been helping the Indonesian government with the protection of Ujung Kulon and the rhinoceroses.

Rarely seen, despite all the interest taken in its welfare, the Javan rhinoceros is similar in size to its Indian relative, but is more lightly built, with a smaller head. The average height at the animal's shoulder is about 140–170 centimetres and its body, including the approximately 70-centimetre-long head, is 305–320 centimetres long. In females, the single horn may be so reduced as to be almost non-existent.

The animal's thick, dark grey skin is deeply folded, with a distinctive 'saddle' on the front of the shoulder, giving it an armoured appearance, similar to its Indian counterpart (*Rhinoceros unicornis*). The hide has an embossed, nodular quality. These characteristics, together with the Javan rhinoceros's virtually hairless hide, easily distinguish it from the smaller, hairier, Sumatran rhinoceros (*Dicerorhinus sumatrensis*). Unlike the

The Javan rhinoceros is a shy and solitary animal which prefers lower forest land to hilly country.

Sumatran species, too, both the Indian and the Javan species have scent glands on their feet. Like the Sumatran, however, the Javan rhinoceros, being a browse-feeder, is armed with more formidable incisors than the grazing African rhinoceros. These are much valued by the Javanese for various magical and medicinal purposes.

The Javan species is not as good a climber as the Sumatran, preferring the lower forest zones along the coast, including swamps, to hilly country. It browses on some 190 plant species from 61 families, ranging from pandanus palms to young bamboos, mangoes, and figs, mostly from secondary growth areas. Like its Sumatran counterpart, the Javan is a retiring, solitary wanderer, although it does not range quite as far afield, travelling a maximum of 15–20 kilometres in a single day. It follows a not always continuous network of trails, often linking wallows and pools or river courses, in which it swims.

As with the Sumatran species, communication among Javan rhinoceroses is mainly olfactory, particularly through urine squirts on vegetation, and communal dung piles, accumulated from the visits of 2–7 animals. Vocal communication is rare. However, the rhinoceros has been observed to 'neigh' (a high blowing whistle), often on scenting a female, 'bleat' (mostly cow–calf communication), snort explosively or even shriek when disturbed, and indulge in lip vibration while feeding, this latter being more 'comfort behaviour' than true communication.

Little is known about the animal's reproductive habits, but the animal's loud trumpet-like roars are probably uttered not just when wounded or fighting, but also during the male's rut. The males conduct an aggressive courtship of the female, following her for some time, and again preface mating with a fight, accompanied by loud roaring. Pregnancy is estimated to last about 16 months and the cow–calf relationship about 2 years.

Little is known of the Javan rhinoceros's life cycle. The species is so shy that sometimes the only physical material for study may be its tracks in forest mud, as viewed here at Ujung Kulon in Java.

Recent WWF studies using track-counts and rhinoceros-triggered photography show that the Ujung Kulon rhinoceroses are breeding successfully.

Indonesia recognizes both the actual and symbolic significance of the Javan rhinoceros's survival. The animal was selected as the logo for the country's major tourism promotion event, 'Visit Indonesia 1991'. The work of the Indonesian government's dedicated Species Conservation chief, Widodo Sukohadi Ramono, in the cause of *in situ* conservation of the rhinoceros in the wild, won him the conservationist Order of the Golden Ark from Prince Bernhard of Holland in late 1991.

After prolonged debate within the wildlife conservationist community in the 1980s, the idea of attempting captive breeding of this fragile animal has been shelved, if not rejected, in favour of *in situ* conservation. The lessons learned from the expensive Sumatran captive-breeding project have been stern, revealing a 30 per cent mortality rate and no captive-breeding in more than five years (partly because the zoos involved did not all establish male–female pairs), statistics that the tiny population of Javan rhinoceros could hardly withstand. One possibility in the future may be the translocation of a few pairs of Javan rhinoceroses to Sumatra's swampy Way Kambas National Park or to Panaitan Island, offshore from Ujung Kulon, to start a second population centre for the animal as a hedge against future disasters. However, there are risks involved in even this much disruption to the rhinoceros, not to mention the expense.

The Green Lady of the Sea

A mermaid goddess has been part of the household of the sultans of Yogyakarta in Java for over 400 years, balancing the powers of heaven and sea with those of the earth, in cosmic union. Kangjeng Ratu Kidul is spirit queen of the sea and also eternal consort to successive generations of sultans of the House of Mataram, teaching them the secrets of government, war, and love. At the ruler's palaces, water-gardens are a dominant theme, in honour of the goddess, and special areas are reserved for the Sultan's regular communions with her. When Sultan Hamengkubuwono X was crowned in 1989, the assembled gathering noted a gust of wind followed by a powerful perfume, and knew that the goddess was still at the Sultan's side.

She is said to particularly love a pale shade of yellow-green, perhaps similar to moonlight, for her outstanding beauty is also said to wax and wane with the monthly phases of the moon. Occasionally, she will drag humans (especially those clad in green swimsuits) down into her kingdom, drowning them, to replenish her supply of subjects, which tends to explain why some Javanese, and the Balinese, are not very enthusiastic swimmers.

Tender Traps

There is a horror-movie, science-fiction kind of resonance to the term 'carnivorous plant', but the pitcher plant is indeed such a beast. More accurately, it is insectivorous. Pitcher plants are often referred to by their Latin name *Nepenthes*, derived from the Greek word for the alcoholic oblivion and banishment of all care reached through the wine goblet.

Basically, all 80 species have specially adapted leaf tips, each with a flip-up lid to prevent too much rainwater from entering the 'monkey's cup' or 'kettle', as the urn-like pitcher is sometimes called. But they take many forms: some are slender appendages on tendrils shooting from trailing vines, others are squat little vats hugging the ground, still others are showy and fleshy with sensuously full, mottled red-and-purple lips. The widest range of species is found in Borneo, home to 30 species; Sumatra has 21, while Java is home to only 2. Generally, pitcher plants are found on poor soils, where their feeding strategy can supplement the lack of nutrients.

Nepenthes sanguinea is just one of about 80 species of pitcher plant, including all those found between the Asian tropics and Australia and New Caledonia, as well as those on Madagascar and the Seychelles. However, it is one of only two found on Java.

The pitcher plant's carnivorous feeding strategy compensates for the nutrient-starved soils which it colonizes. Here a fly falls victim to the plant. Once an insect has fallen into the pitcher, it is digested by the plant's special enzymes.

The sensual colours and smells of the plants attract inquisitive insects, which then lose their footing and fall into the pitcher. Inside the pitcher are secreted digestive enzymes which then consume the plant's prey, drawing from them nutritious nitrogen, phosphorus, and mineral salts. Several species of insects—some spiders, tadpoles, mosquito larvae, and tiny crabs, for instance—have developed a resistance to these enzymes, and so can live in the pitchers, benefiting from the bonus foods which fall in.

As a last resort, the dubious liquid inside a pitcher plant could be a boon to the thirsty hiker; nineteenth-century naturalist Alfred Russel Wallace said it was 'very palatable, though rather warm', despite its uninviting appearance. The largest species, *Nepenthes rajah*, can easily hold 2.27 litres of fluid. The pitchers have also been put to other uses in village life: the liquid inside used to bathe eyes and soothe both inflamed skin and coughs, the crushed leaves and roots applied as a medicinal astringent, the roots boiled for stomach ailments, the stems used for rope-weaving fibres, and the pitchers themselves occasionally used for cooking glutinous rice, by all accounts imparting a delicate flavour and colour to the rice.

The Spotted One

Agile and swift, the leopard (*Panthera pardus*) is perhaps the most adaptable of all the big cats, which may explain why it has survived so well in Java; recent WWF sightings by hidden cameras in the Ujung Kulon National Park suggested a local population of as many as 60 leopards.

The leopard can climb, run, and swim, and is an accomplished hunter, chiefly of pigs and other small mammals such as mouse deer (*Tragulus* spp.), monkeys, or binturongs (*Arctictis binturong*), but its catholic diet can extend to delicacies such as crabs, cattle, or dogs. Prey is usually devoured from the entrails on, and leftovers sometimes stored high up on tree branches.

The Javan leopard is a distinct subspecies, but the so-called black panther is in fact what scientists call a melanistic version of the yellow-and-black rosetted leopard. Close scrutiny of a panther's coat reveals similar rosettes. The aquamarine eyes of the leopard manifest the same mysterious inscrutability as the domestic cat's, perhaps more so than in the tiger. Having once weathered a playful attack by a semi-wild, newly rehabilitated specimen of its Indian relative, this writer can testify to the animal's brooding unpredictability.

Java has the leopard (*Panthera pardus*), while Sumatra and Kalimantan do not. Leopards are good climbers and will hunt anything from crabs to pigs and monkeys. They are still common in the Ujung Kulon National Park, as well as at Baluran and Meru Betiri on Java.

In 1984, zoologists led by Bas E. van Helvoort were astounded to discover a new population of leopards on Kangean Island 200 kilometres due east of Madura and due north of Bali. The leopard's presence on Java is something of a mystery, since there are no leopards on Sumatra or Borneo. The island was once used as a place of exile or penal colony, both by the traditional sultans and by the Dutch colonial rulers, so it could be that Javan leopards were deliberately introduced on to the island to deter would-be escapees.

Leopards, quite unlike tigers, are generally at home in the trees and will often store their kills high up on tree branches. They are night hunters, stalking pigs and deer.

At only 10 days old, these leopard cubs sleep a lot. The black animal, a normal leopard, is what scientists call a melanistic version—the typical spotted pattern is there, but is hard to see through the black colour.

6 Kalimantan: River of Diamonds

'BORNEO' is a word with great emotive appeal to the Western psyche, but few realize that Kalimantan covers two-thirds of this, the world's third biggest island. Kalimantan, once Dutch Borneo, makes up more than a quarter of Indonesia's territory, yet holds only 5 per cent of the country's population. It boasts Indonesia's longest rivers: the Kapuas (1143 kilometres), the Mahakam (650 kilometres), and the Barito (890 kilometres).

Clad in rugged rain forest, traced by mighty rivers, and edged by extensive mangrove swamps, Kalimantan was for a long time a territory favoured only by the fearsome Dayak head-hunting natives, its rivers their roads, its mountains the abode of their dead souls awaiting rebirth. 'Dayak' is a general term for various native peoples, including the once-nomadic Punan, Iban, Kelabit, Kayan, Kenyah, and Land Dayak.

◁
Hornbills are revered by the Dayak peoples of Kalimantan. The bird's plumage is treasured for ceremonial dances such as the one this dancer is about to perform. Some dances also depict the hornbill as a spirit-being, symbolic of resurrection after death.

Kalimantan is a province criss-crossed by waterways; it boasts Indonesia's longest rivers.

KALIMANTAN

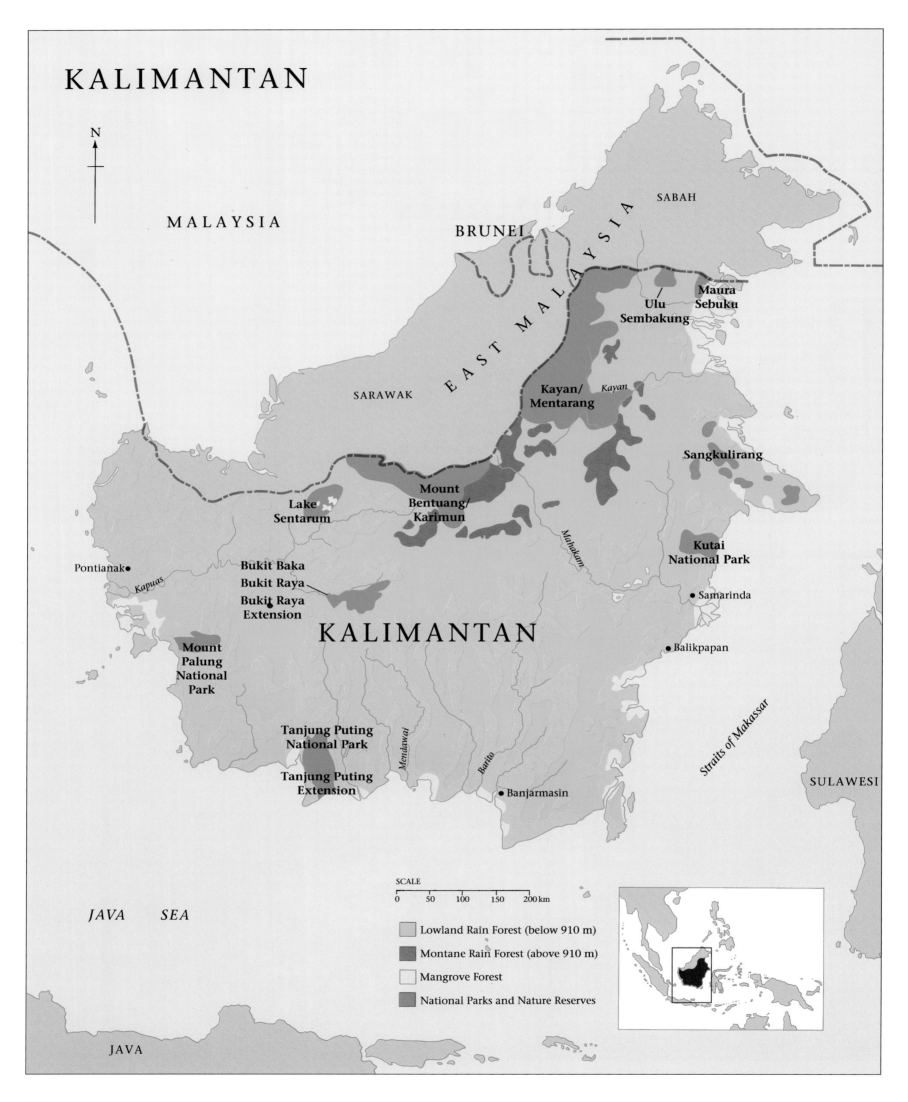

MALAYSIA

SABAH

BRUNEI

E A S T   M A L A Y S I A

Ulu
Sembakung

Maura
Sebuku

SARAWAK

Kayan/
Mentarang

Kayan

Sangkulirang

Mount
Bentuang/
Karimun

Lake
Sentarum

Pontianak

Kapuas

Kutai
National Park

Bukit Baka
Bukit Raya
Bukit Raya
Extension

• Samarinda

KALIMANTAN

Mahakam

Mount
Palung
National
Park

• Balikpapan

Tanjung Puting
National Park

Mendawai

Barito

Straits of Makassar

Tanjung Puting
Extension

SULAWESI

• Banjarmasin

JAVA   SEA

SCALE

0   50   100   150   200 km

Lowland Rain Forest (below 910 m)

Montane Rain Forest (above 910 m)

Mangrove Forest

National Parks and Nature Reserves

JAVA

While scattered Hindu statues, inscriptions, and other relics in Kalimantan point to ancient ties with Indian civilization, huge Chinese ceramic jars standing in some of the Dayaks' communal longhouses attest to the long-standing trading relationship between Borneo and China, dating from at least the seventh century AD. Borneo's resource-rich interior had much to offer the Chinese, from gold and diamonds to rattan and gutta-percha latex, peppercorns, edible bird's nests, hornbill feathers, rhinoceros horns, even semi-magical medicinal 'bezoar' stones found in monkeys' gall bladders, and incense-bearing woods and resins. In more recent times, the most sought-after resource in Kalimantan, particularly in the east coast area around the town of Balikpapan, has been oil.

However, Kalimantan's most precious resource is its forest. Here, more than anywhere else in Indonesia, can be seen the full glory of the tropical rain forest as the West has always imagined it. In the forests of Borneo, that bald scientific term 'biodiversity' assumes meaning amid 11,000 species of flowering plant, a third of them endemic. There are perhaps up to 4,000 tree species in Kalimantan, with about 150 different species of tree for every 1 hectare of forest, many of them tall dipterocarp hardwoods. Approximately 60 per cent of Kalimantan is still forested, albeit partly logged, and what is of the utmost significance is the fact that the forest exists in large continuous blocks, unlike the fragmented pockets typical of Sumatra.

Soaring trees, tangled lianas, thorny rattans, strangling figs (*Ficus* spp.), cascades of delicate epiphytic orchids, and the grotesque *Rafflesia* flower are among the myriad colourful threads that combine to form the exotic tapestry of Kalimantan's rain forest, bathed annually by 4500 millimetres of rain. Kalimantan is also an important centre of origin for tropical fruit

The rain forest is perhaps at its most glorious in Kalimantan of all the Indonesian rain forest locations; here, there are about 150 different species of tree for every 1 hectare of forest.

According to the US National Academy of Sciences (NAS) in 1980, a 'typical' 4-square-mile patch of rainforest contains up to 1,500 species of flowering plants, 750 species of trees, 125 of mammals, 400 of birds, 100 of reptiles, 60 of amphibians and 150 of butterflies.

(Charles Secrett of the 'Friends of the Earth' forest protection group, 1986)

The black orchid (*Coelogyne pandurata*) is a strange green orchid with black 'warts' on its lip. The forests of tropical South-East Asia have the lion's share of the world's approximately 17,500 species of orchid.

trees such as the mango (*Mangifera indica*), durian (*Durio zibethinus*), and breadfruit (*Artocarpus altilis*). There are no fewer than 16 species of mango in East Kalimantan, 3 of them inedible.

Kalimantan's wildlife is no less varied. There are about 479 species of bird, 1 endemic (but 35 endemic to Borneo), ranging from the splendid peacock-tailed great Argus pheasant (*Argusianus argus*) strutting his forest-floor stage while courting his lady-love, to tiny birds such as the pitta. The endemics include the small white-fronted falconet (*Microhierax latifrons*), the red-breasted wood partridge (*Arborophila hyperythra*), and the glittering blue-wattled Bulwer's pheasant (*Lophura bulweri*) with its snow-white tail and red-and-violet neck feathers. The great Argus pheasant's feathers, which forest peoples use to adorn their ceremonial head-dresses, are very much in demand, unfortunately for the bird. Other birds found in the Kalimantan forests include the crested fireback (*Lophura ignita*) and the endemic Bornean peacock-pheasant (*Polyplectron malacense schleiermacheri*).

Birds play a special role in traditional Dayak culture as intermediaries with the gods and spirits, and as augurs of good and bad omens. The direction of a bird's flight—to the left or right, away from the spectator, or in a diving motion—has relevance to the decisions of the day, particularly with reference to the various cycles of the rice harvest, or when hunting, in the case of the nomadic forest Punans. The Brahminy kite (*Haliastur indus*), in particular, is considered to be a reincarnation of a deity much revered by the Iban Dayaks, Singalong Burong, but many other birds are also regarded as ritually important, the maroon woodpecker (*Blythipicus rubiginosus*) and the banded kingfisher (*Lacedo pulchella*) among them. Many Bornean peoples can imitate bird and animal sounds very accurately, a skill they use to attract their prey when hunting. Naturalist Charles Hose noted in the 1920s how the Kayans used a short length of bamboo to imitate bird sounds. He had seen an owlet so deceived by this that it flew into the native hut whence the sound had come, looking for its interlocutor.

This breadfruit (*Artocarpus* sp.) is among many fruit trees indigenous to Kalimantan. The Kalimantan rain forest is an important gene pool for trees of potentially high commercial and food value.

Kalimantan rejoices in several exotic forest pheasants, one of which is this crested wood partridge (*Rollulus rouloul*), which wears a flamboyant red crest with red feet to match.

With a crest that looks as though it has been tied on for decoration only, the crested fireback (*Lophura ignita*) is another bizarre ground-dwelling bird in the Kalimantan forests.

> We cannot be surprised that this mysticism of nature should induce in the native a spirit of credulity and superstition. He learns to see the play of spirits in every natural phenomenon—the crackle of a dry bough, the monotonous note of the cicada's nightly song, or the shrill morning signals of the Argus pheasant. He becomes an incurable animist, whose attitude to Nature in all her phenomena is one of tactful consideration. He treats the soul of all things with gentle deference. If gnawing hunger compels him to lay hands on a pith-filled palm, he prays and sacrifices first to its soul.
>
> (Eric Mjöberg, *Forest Life and Adventures in the Malay Archipelago*, 1930)

The peacock-pheasant (*Polyplectron malacense*) is found in Kalimantan as the endemic subspecies *Polyplectron malacense schleiermacheri*.

Of Kalimantan's 221 land mammals, about 28 species are endemic. To each Punan tribesman, one of these animals is a personal totem whose flesh he cannot eat; all species have their Punan protector since each tribesman is assigned a different creature as his own. The sun bear or honey bear (*Helarctos malayanus*) is particularly respected, and perhaps wisely, left alone. The wide-eyed slow loris (*Nycticebus coucang*), whose powdered liver the Dayaks believe is a powerful aphrodisiac calculated to seduce the female of their choice, and the swivel-headed Western tarsier (*Tarsius bancanus*), both nocturnal primates, live here, alongside a plethora of squirrels—all coming under the umbrella local term, *tupai*—including the common giant squirrel (*Ratufa affinis*), the elegant chestnut-and-black marked Prevost's squirrel (*Callosciurus prevostii*), the common plantain (*Callosciurus notatus*) and slender squirrels (*Sundasciurus tenuis*), the tufted ground squirrel (*Rheithrosciurus macrotis*), the flying squirrel (at least 14 species), and the tiny pygmy squirrel (*Exilisciurus* spp.).

Gibbons, such as the agile or black-handed gibbon (*Hylobates agilis*), and orang-utans (*Pongo pygmaeus*) share a treetop niche, searching for fruit and figs, while five species of leaf monkey (*Presbytis* spp.) at the same level stick mainly to foliage. These leaf-eaters have large stomachs much like those of cows and deer, divided into several compartments, to ferment, break down, and neutralize toxic vegetable fodder which may weigh around a quarter of the monkey's own body weight. While these primates' acrobatic antics are impressive, perhaps even more remarkable are creatures such as lizards, frogs, and snakes which have developed the ability to 'fly', or at least glide. Flying squirrels and flying lemurs (*Cynocephalus variegatus*) can glide for as far as 100 metres, stretching a flap of skin between their limbs to resemble a tautly tensed kite. The squirrels can even steer a sort of slalom course between the trees. In this way, such animals avoid forest-floor enemies and make forest travel less exhausting. The flying lemur is otherwise somewhat dull, spending much of its time seemingly frozen to tree-trunks.

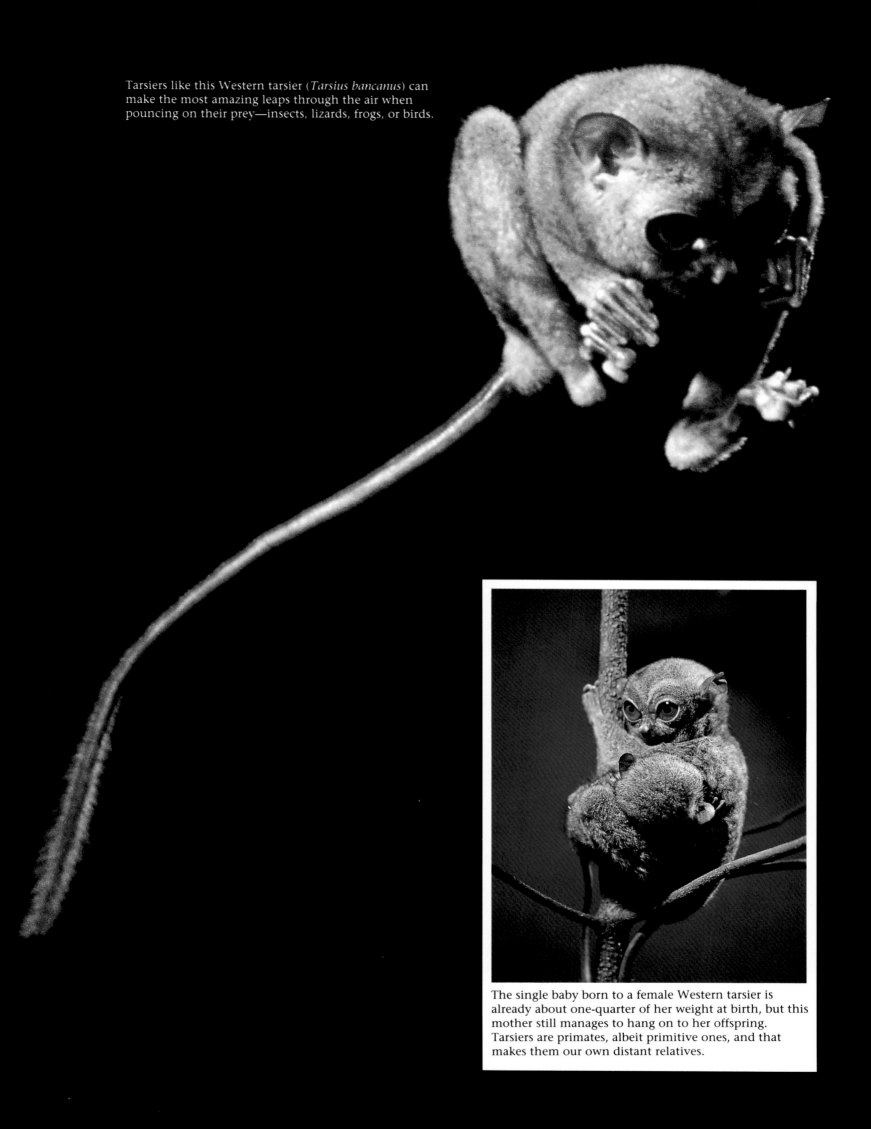

Tarsiers like this Western tarsier (*Tarsius bancanus*) can make the most amazing leaps through the air when pouncing on their prey—insects, lizards, frogs, or birds.

The single baby born to a female Western tarsier is already about one-quarter of her weight at birth, but this mother still manages to hang on to her offspring. Tarsiers are primates, albeit primitive ones, and that makes them our own distant relatives.

The Bornean gibbon (*Hylobates muelleri*), which is endemic to Borneo, is found north of the Kapuas–Barito river drainage, while the similar-looking agile gibbon (*Hylobates agilis*) lives south of that area. In the zone where the two species overlap, they sometimes mate.

Across the forest floor trundles that quaintly armoured scaly anteater, the sticky-tongued pangolin (*Manis javanica*). Some say that after the pangolin has broken open an ant nest for dinner, sometimes climbing a tree to do so, it will carefully close its scales to trap the ants swarming over its body, then take a swim and deliberately open up its scales underwater so that the ants will float out at convenient tongue's-length for it to savour. The pangolin can consume literally hundreds of thousands of ants or termites in a day. Snuffling the floor, too, is the bearded pig (*Sus barbatus*). One of the great spectacles of the Kalimantan forest is the seasonal migration of thousands of these pigs in search of fruit, a source of great delight to marauding Dayak and Punan hunters, who relish their flesh.

A toothless wonder, the pangolin (*Manis javanica*) attacks termite nests with its claws and long sticky tongue. Contrary to appearances, the pangolin, also known as the scaly anteater, is a mammal. The scaly animal can roll itself up into a smoothly armoured ball if alarmed.

The reticulated python (*Python reticulatus*) is an impressive snake reaching great lengths, often well over 4 metres and sometimes as much as 9 metres, but it is not particularly dangerous to human beings. As this photo shows, python mothers can be as maternal as any others.

Snakes are common in the Bornean forest, as they are throughout the Indonesian forest, but not so easily encountered as first-timers may fear, for they prefer to get out of the way when they sense the vibrations of human feet treading the ground; they feel rather than hear. Among the snakes found in Kalimantan are the beautifully named and even more beautiful-to-behold rainbow-coloured paradise tree snake (*Chrysopelea paradisi*), an accomplished glider, and the sinuous whip snake (*Ahaetulla* spp.), both relatively harmless.

In general, the brighter a snake's markings, the more one should beware, although this does not apply to one of the most dangerous of snakes: the common black spitting cobra (*Naja naja*). A cobra, however, will warn intruders off first by spreading its hood and raising itself erect. Its objective is to scare the enemy away, and the commonest species will more likely spit venom into its tormentor's eyes before biting, blinding temporarily or permanently, depending on the speed with which treatment is sought. The more aggressive and much larger (as much as 4 metres long), deadly king cobra (*Ophiophagus hannah*), however, is a much more daunting proposition, whose bite is likely to be lethal. The reticulated python (*Python reticulatus*), with its attractively marbled markings, is extremely common, impressively massive, and kills by squeezing and crushing. Yet python attacks on human beings are so rare as to be insignificant.

No tigers or leopards inhabit the Bornean forest, but there is the handsomely marked clouded leopard (*Neofelis nebulosa*), probably watching from a perch up in the trees. Known as the 'tree leopard' (or more literally translated, 'branch leopard') by the local peoples, this beast is much revered, and a special cleansing ritual must be undergone if one is killed. Much more dangerous in any encounter with humans, however, is the white-bibbed sun bear, also a resident of the Kalimantan forest; its vicious claws are designed for climbing and tearing trees open to get at grubs and bees' nests. Many indigenous names for animals are essentially onomatopoeic; *bruang*, in imitation of the bear's roar, is interestingly close to the Western 'Bruin'. Not exactly a dedicated carnivore, the sun bear is more of an omnivore, and seems immune to bee stings as it rummages for forest honey. Partly because it is very shortsighted, the bear is easily panicked and may attack humans without warning, rearing up on to its hind legs. This presents a very real threat to forest peoples who may be after the same honey as the bear. However, in one incident known to this writer, a towel flapped in the bear's eyes at the very last moment of its attack

The clouded leopard (*Neofelis nebulosa*) of Kalimantan spends rather more of its time in trees than the leopard (*Panthera pardus*) of Java. It is greatly revered by the native peoples of Kalimantan. None the less, the cat's beautifully marked coat is much sought after for ceremonial costumes.

served to confuse and deter the animal, and it turned aside, lumbering off into the undergrowth.

Unfortunately, humans do not share the sun bear's immunity to bee attacks. Hornet stings are by far the most serious risk to the forest hiker's safety, more so even than the sun bear or tiger. Their repetitive stings delivered *en masse* can produce kidney collapse, extreme shock, and suffocation leading to death or permanent disability. The forest bee is very wild; there is no comparable history of domestication, as in Europe since Roman times, although experiments have been made recently in Sumatra. Hence, the forest hiker's code contains this important injunction: never idly poke or prod anonymous lumps of leaves, twigs, or clay, whether on the forest floor or suspended above, and walk carefully and quietly at all times.

The full panoply of Bornean wildlife, including less alarming carnivores than the sun bear, such as the much smaller leopard cat (*Prionailurus bengalensis*), otter (Mustelidae), civet (Viverridae), and mongoose (Herpestidae), though more easily observed in the accessible national parks of Tanjung Puting in the south, and Kutai in East Kalimantan, can also be found in inaccessible wild regions like the Kayan–Mentarang Reserve, and the Mount Bentuang and Karimun regions, both hilly areas on East Malaysia's borders. Another wildlife-watching location is the Bukit Raya/Bukit Baka area of Central and West Kalimantan. One of the problems of a politically divided area like Borneo, which contains the East Malaysian states of Sarawak and Sabah, and the small oil-rich sultanate of Brunei, besides Kalimantan, is that Nature rarely respects national borders. The need for trans-border parks jointly run by Malaysia and Indonesia is often felt, and happily, the two countries have already made some progress in this area.

The sun bear (*Helarctos malayanus*), also known as the honey bear, is possibly the most dangerous animal in the Kalimantan rain forest. The bear has a well-deserved reputation for bad temper. Armed with vicious claws designed to rip up tree bark and termites' nests, as well as beehives, it makes a formidable enemy.

It looks like a crocodile, but it is only a relative of the crocodile, the fish-eating false gavial (*Tomistoma schlegeli*). This shy beast lurks for long periods below the surface of rivers and swamps and may even sleep underwater, on the river-bed.

The Bentuang–Karimun area on the Malaysian Sarawak state border in West Kalimantan, identified by some experts as an area of global conservation significance, is one example, linking as it does with Malaysia's Lanjak Entimau Reserve. The combined forest of these two reserves protects orang-utans, gibbons (called *wah-wah-nuk* by forest peoples, for their plaintive song), clouded leopards, banteng (*Bos javanicus*), and countless rare plants. There are plans for similar co-operation on the large, 16 000 square-kilometre Kayan–Mentarang park and the adjacent 5000-square-kilometre Ulu Sembakung area in East Kalimantan, lying along the Malaysian Sarawak and Sabah state borders. This region is botanically very rich and houses many endemic mammals and birds, as well as 3 endemic swallowtail butterflies (Papilionidae). There probably are elephants (*Elephas maximus*) and rhinoceroses in some unsurveyed primary forest within the Ulu Sembakung reserve. Certainly, Kayan–Mentarang adjoins Sarawak's Pulong Tau reserve area, which is known to harbour one of the last remaining Sumatran rhinoceros (*Dicerorhinus sumatrensis*) populations on Borneo. Such projects could be a model for bilateral co-operation on conservation within the ASEAN group of countries (Indonesia, Malaysia, Singapore, Thailand, the Philippines, and Brunei).

The Tanjung Puting park covers an area of just over 3000 square kilometres, much of it dry heath forest or peat swamp forest. Besides being home to the grotesque proboscis monkey (*Nasalis larvatus*), gibbons, the clouded leopard, several deer, and the small fish-eating false gavial crocodile (*Tomistoma schlegeli*), it is also the location for an orang-utan rehabilitation centre, Camp Leakey Research Station. Encounters between crab-eating long-tailed macaques (*Macaca fascicularis*) and crocodiles (*Crocodylus porosus*) lurking beneath the mud at the river's edge offer fascinating studies of crocodilian behaviour. Once it sees its opportunity, the crocodile will knock several monkeys aside with a stunning blow from its tail, then carry them off to be stored under a boulder until they are suitably decayed and fit for a crocodile feast, rather like the semi-rotting British or Dutch hung pheasant.

Established in 1936 by the local sultan, Kutai, at 2000 square kilometres, is only slightly smaller than Tanjung Puting but bears the scars of logging and of devastation by fire at the end of 1982, just after being declared a national park. The park represents an important rain forest reserve, and also has one of the last remaining ironwood forests once typical of East Kalimantan.

The phenomenon of fire in what should be evergreen moist forest is an indicator of changing world climate patterns, and of a dangerous dry-out in the forest. The 1982 blaze ripped through 35 000 square kilometres of forest, besides another 15 000 square kilometres in North Borneo, destroying not only wildlife but also an estimated US$8 billion worth of potential commercial timber. The resulting smoke was enough to affect visibility, and consequently plane schedules, as far away as Malaysia and Singapore. There have been other such fires since, both in Kalimantan and in Sumatra, notably in late 1989, again in late 1991, and in late 1992. Some think—and satellite imagery bears this out—that the fire smoulders on permanently, somewhere deep beneath the peat soils and coal seams of the Kalimantan forest, merely biding its time to break out once more.

It is most likely that the great 1982–3 fire resulted from a tragic convergence of factors such as logging clearance and kindling-like detritus left behind, unusually dry weather conditions provoked by the notorious 'El Nino Current', pressures on the land arising from transmigration settlement, and the Dayaks' traditional slash-and-burn *ladang* style of cultivation—a technique which is usually quite acceptable when population pressures are not so high as to prevent a reasonable fallow period for cultivated forest soils to recover.

Kutai now offers an unusual opportunity for biologists to study the damage and monitor the pace of regeneration in this damaged park, which is believed to harbour banteng, deer, sun bears, and small carnivores, besides about 300 species of bird, including 8 species of hornbill. Although Kutai featured on the International Union for the Conservation of Nature and Natural Resources' (IUCN) list of the world's most threatened protected areas in 1984, there is hope for the future. Several industrial concerns benefiting from East Kalimantan's rich natural resources have already offered assistance for the preservation and rehabilitation of Kutai. It seems only right that some of the profits derived from Kalimantan's natural wealth should be ploughed back into the region's most precious resource— the natural forest.

Unidentified resin-eating insect on a resin-bearing tree-trunk in Tanjung Puting National Park. The creature's camouflage is striking, its whole body resembling the shiny resin it is devouring.

Monkey in the Water

'Ludicrous' was nineteenth-century colonial officer and naturalist Charles Hose's considered opinion of this poor monkey's appearance. Rather rudely, some Borneans call him 'the Dutchman monkey'. The proboscis monkey (*Nasalis larvatus*), found only on Borneo, has two notable characteristics: a comically large, long, and pendulous red nose, and an extraordinary penchant for frolicking in water. It is not clear what advantage the proboscis monkey gains from its strange nasal appendage. It even has to push it aside to feed properly. However, some zoologists say it is simply attractive to female proboscis monkeys.

A large monkey, weighing as much as 24 kilograms, the proboscis monkey has reddish hair, an off-white tail, and something of a pot-belly. It is found chiefly in tall coastal and riverine forests, rarely inland, where it hunts exclusively for leaves and the harder, starchier fruits, never digressing to any form of animal protein. It has a highly complex stomach, built to cope with this diet. This dependence on very specific habitats and foods makes the proboscis monkey a species particularly vulnerable to any sudden change in its environment, and almost impossible to keep in captivity.

A typical proboscis monkey family comprises a male and a harem of 3–4 females, plus babies, who usually move around together in groups of between 10 and 40. Males constantly show off, crashing through the trees and honking loudly, to attract new females. Some scientists believe that the monkey's nose might be a resonating chamber for magnifying these calls, since the nose is far more developed in the males.

Supremely and unusually at home in the water, the monkeys dive in, splash around, and even swim underwater for short periods. They are helped in this by their partially webbed back feet, although crocodiles in the rivers they frequent must surely constitute a peril to them. There are even records of these monkeys being caught in fishermen's nets as far as 3 kilometres offshore. A good place to observe these extraordinary monkeys is on Kaget Island, two islands in the Barito River, just south of Banjarmasin, and also at the Mount Palung and Tanjung Puting national parks.

Proboscis monkeys or 'Dutchman monkeys' (*Nasalis larvatus*), found only in Borneo, perch peacefully at the top of trees, their long and sturdy grey tails hanging down. Twilight is upon them and each family will soon settle into its regular sleeping spot.

The proboscis monkey is quite at ease in the water. This one peeks comically through the mangrove vegetation, its pendulous nose just resting on the surface of the water, at Tanjung Puting National Park.

The noses of female and young proboscis monkeys are not as noticeable as those of the males; they are more pointed or snubbed, as seen here. It is thought that the male's droopy nose may be sexually attractive to the females.

A mature male proboscis monkey like this may weigh more than 20 kilograms, a weight achieved solely on a diet of tough mangrove leaves. He usually manages a harem of about 3–4 females. The distended-looking belly is normal in an animal which has to process masses of vegetation at a time.

Wards of the Space Bird

The Dayak peoples of Kalimantan believe they originally came from another planet. Their ancestor, Raja Bunu, still lives there, with his two brothers, Raja Sangin and Raja Sangian, all of them the children of Manyamai Tunggul Garing, the God of Gods. One day, Manyamai Tunggul Garing told Raja Bunu to go and settle Earth, so Raja Bunu left in a special vehicle made of gold and landed in Central Kalimantan at a place which today is called Bukit Raya.

Raja Bunu's companion on this epic journey was the *enggang* bird—the helmeted hornbill (*Rhinoplax vigil*)—ever since regarded by the Dayaks as a sacred bird. No Dayak ceremony proceeds without the prerequisite blessing to seek protection from the *enggang* bird, whose effigy is carved on a tall wooden pole erected at the ceremonial grounds.

> I look them in the eyes, and the eyes that look back at mine are exactly the same. It is so easy to relate to them. There is a total absence of malice about them.... I think orang-utans preserve an innocence we lost when we left the Garden of Eden. Orang-utans never left it, never made the break.... That's why we perceive the basic goodness in them.
>
> (Birute Galdikas, orang-utan rehabilitation expert)

Their babies cry like ours, play like ours. They make love like us. Their dexterous fingers grip objects in the same way ours do. They use tools. They are intelligent enough to learn new behaviour. They share 97 per cent of their genetic material with us. It is very, very hard to resist anthropomorphizing the great ape, the orang-utan or 'man of the forest', delightfully named *Pongo pygmaeus*. Perhaps the biggest difference between us and the orang-utan, apart from its shaggy coat of reddish hair, is its enormous strength, about four times our own. Weighing up to 90 kilograms and standing up to 140 centimetres tall, it can take the husk off a coconut with its bare hands. Its arms may be as long as 2 metres and it is this feature that tends to convey the impression that the animal is much bigger than it really is.

Only Sumatra and Borneo are home to these splendid animals. Thus, Indonesia is burdened with much of the responsibility for their survival, since there are only a few thousand of the apes left in the wild, and their numbers are on the decline.

Adult males are often fearsome of visage, adorned with the puffy cheek flanges, hairy beards, and shaggy long-hair coats typical of the Kalimantan animal. They are highly competitive, avoiding contact with one another, and roaring long loud calls to warn each other off, since encounters can only result in violent combat. Actual orang-utan battles are rare, but when they do occur, they usually leave the antagonists scarred with bites and

The male Bornean orang-utan (*Pongo pygmaeus*) is an imposing sight, with his dark cheek flanges and powerful arms. A full-grown adult male may stand about 140 centimetres tall and weigh as much as 90 kilograms. The Bornean animal is a subspecies quite separate from the Sumatran subspecies.

other injuries. Fortunately, however, it seems one male can easily establish his dominance over another, merely through the medium of his call.

Primatologist John MacKinnon, who has done seminal work on this great ape, has reported that the orang-utan is both quiet and shy in its natural habitat, to the chagrin of many a naturalist forced to trek miles through tough territory in search of it. When found, it then turns out to be remarkably inactive most of the time. In this, the brooding orang-utan offers sharp contrast with the more boisterous Bornean gibbon (*Hylobates muelleri*). Largely vegetarian and frugivorous, the orang-utan will also snack on insects, lizards, young birds, and eggs. In fact, to maintain its huge bulk, it has to eat, and eat, and eat. Its favourite foods are human favourites too: fig, rambutan, banana, and durian. It seems to have an uncanny knowledge of exactly when and where there will be fruiting trees, for which it heads with great precision, an amazing achievement in forests where there may be some 3,000 species of tree, most with irregular fruiting seasons. The orang-utan seems to carry around a mental map which, coupled with high intelligence and a well-developed memory, helps it find the fruit it wants. The best way to find fruit in a Bornean forest is to follow an orang-utan.

> What I discovered with the orang-utan was … a completely new variant on the apely theme: an unlikely hotchpotch of spare parts, with a chimpanzee's body but gibbon's long arms, the gorilla's sexual dimorphism but its own uniquely solitary habits; all within the framework of a strangely man-like society.
>
> (John MacKinnon, *The Ape Within Us*, 1978)

Powerful yet often nimble and gentle, the hands of an orang-utan evoke wonder for their similarity to our own. These are the same hands which can husk a coconut and, if hearsay is to be believed, easily rip a marauding crocodile or a python apart.

Proud and protective motherhood is written all over the face of this female orang-utan. It is one of the great tragedies of Man's dialogue with this closely related species that, because the baby human-like orang-utan is so appealing, the mother is often shot dead in order to procure the screaming baby for an illegal pet market.

Every night, the orang-utan devotes about half an hour to making a new 'bed' or nest of leafy twigs for itself. The animal spends much of its time in the trees, swinging laboriously through the canopy owing to its great weight, but very occasionally walks the forest floor on two legs, like a man.

Orang-utans are not essentially social animals, although MacKinnon reports that they seem more sociable in Sumatra than in Borneo, possibly in order to offer each other better defence against tigers, which do not exist in Borneo. They prefer to live alone or in mother–child duos for up to six years. Male and female orang-utans do not form enduring couples, but as has been hinted already, their somewhat infrequent couplings are both prolonged and mutually erotic, or even acrobatic.

The stronger males frequently rough-handle frightened, screaming females in the process. Rape of females is not unknown, although the first to make an amorous or sexual advance may be either the male or the female. Other human-like sexual activity includes masturbation, in both sexes, and 'homosexuality'. Like humans again, orang-utan females are technically sexually receptive at any time of the year once they have reached maturity, at about 10 years of age; they do not come 'on heat' as some other primates do, although they are fertile for only a few days of their 30-day cycle.

Orang-utan pregnancies last about nine months, and single babies are the usual rule, just as in humans. As with humans, too, the young are completely helpless without their mothers for the crucial early years of life. The young orang-utan is extremely inquisitive, and not a little mischievous. This writer once made the mistake of leaving her washed shirt on the line overnight at an orang-utan rehabilitation station. On finding it missing the following morning, she took a short trek into the forest and discovered a dozing junior orang-utan on the forest floor, clutching the apparently desirable shirt to its bosom. It took some persuasion to remove the garment from its grip. John MacKinnon has

Orang-utans make nests of vegetation every night to sleep in, high up in the trees. This young orang-utan is getting ready for a night's rest.

noted that young orang-utans indulge in exuberantly happy play together once they meet up, but such meetings are rare. Most orang-utan babies grow up isolated, becoming, as a result, largely anti-social adults.

This isolation means, of course, that the baby orang-utan's world focuses heavily on its mother. It is more than usually upset, therefore, when its mother gives birth to a new and rival baby. Births normally occur every three or four years. For the first time, baby number one is banished from its mother's night-nest, and asked to make its own separate little bed just a few metres above hers. Still, it will stay with its mother for another year or so before attempting to survive alone. Many youngsters do not completely leave their mother until they reach sexual maturity, at about 7 years old. Their life expectancy in the wild is believed to be about 30 years.

But why is this ape so uniquely solitary? MacKinnon says that the answer lies in the distribution and availability of the animal's main food, forest fruit. A slow-moving orang-utan group can only find so much fruit in a day and it would have to be spread thin throughout the whole group.

The orang-utan's cuteness has misled many into trading them as pets, usually killing the mother to obtain the screaming baby. There are an estimated 1,000 such pets in Taiwan today. Keeping orang-utans as pets is now illegal in Indonesia, but a black-market trade continues. This explains the need for rehabilitation programmes at centres such as Tanjung Puting in Kalimantan, where as many as 50 'pupil' orang-utans may be under the wing of the centre's well-known director, Canadian-born Dr Birute Galdikas, at any one time, and at Bohorok in Sumatra. Dr Galdikas, originally a protégé of the famous Africa expert, Dr Louis Leakey, after whom her camp is named, has dedicated more than 20 years of her life to the study and protection of orang-utans.

The objective of such programmes is to nurse these orphans and teach them, step by step, as their mothers would have done, how to survive in the forest. Food is gradually withheld from the growing apes and human contact progressively minimized so that, eventually, they will go their own way in the wild. Whether or not such rehabilitation schemes have any real value is still fiercely debated within the scientific world. It is important to keep the orang-utan pupils away from too much human contact, something these centres have found hard to achieve, owing to over-enthusiastic tourists and visitors.

The biggest risk of all is that the rehabilitated orang-utan may transport the diseases of 'civilization' into healthy wild populations, and also that they may unnaturally swell the orang-utan population beyond the carrying capacity of the area where they are released. Whether a released orang-utan can really survive in the wild is another moot point, as is the possibility that they may now be so familiar with human beings as to have bred contempt for us, and so pose a danger to visitors or forest trekkers. However, most experts believe rehabilitation centres more than make up for these risks with the close-range education they give the public on why these animals are worth protecting.

Man, unfortunately, has been in conflict with these marvellous apes for a very long time. There is concrete, archaeological evidence that prehistoric South-East Asian men routinely hunted the orang-utan for food. Yet, the relatively gentle orang-utan will rarely do more than hail broken twigs and branches on to the heads of its tormentors, from its perch on high, close to the forest canopy. Persistent tales of orang-utans wantonly attacking, or even raping, humans have never been authenticated to any independent observer's satisfaction.

The orang-utan can probably survive careful, selective logging, particularly the kind that leaves fig and other fruit trees intact, but the animal does require reasonable areas of gently sloping hilly forest since it is rarely found at heights over 500 metres above sea-level. As one of our nearest relatives, the orang-utan deserves our protection.

In nature, the orang-utan is a solitary rather than a social animal. However, females do move around with their young for some time. Orang-utans are dependent on their mothers for the first two years and remain with them for up to six years. Males do not take any responsibility for child care.

Evidence of a culture which has long known the sophistication of metal-smelting, the Kalimantan blow-gun is a particularly efficient instrument of death vital to the survival of the forest hunter. (Apa Photo)

Deadly Messenger

The Kalimantan blow-gun or *sumpitan* is one of the most efficient known, a long, straight, hollowed shaft of strong ironwood tipped with a sharp spear-blade, unlike the flimsy bamboo tube blowpipes found among other South-East Asian forest dwellers. The unusual bayonet-style blade attachment, and the neat boring of the blow-gun itself, are the legacy of iron-smelting technology which first arrived in Borneo in about the sixth century AD. It is interesting to compare the bore-hole in the blow-gun, effected by a metal drill, with the more primitive method adopted elsewhere: this necessitated splitting the bamboo, grooving a channel inside, and then rebinding the two halves of the bamboo shaft together.

The 2- or 3-metre-long Bornean blow-gun can deliver a 25-centimetre poisoned dart with deadly accuracy over an astonishing 100 metres. When confronted with a dangerous or angry wounded animal at close quarters—a 100-kilogram wild boar or a raging sun bear—the blow-gun bayonet can be used to finish it off. The strychnine-related poisons tipping the blow-gun dart, known generically as *ipoh* to the Malay and *tajam* to the Punans, coming from the upas tree, are typical non-timber forest products, deriving from plants such as the *Antiaris toxicaria* tree (from the sap) or the *Derris elliptica* shrub (from the roots, also known as *tuba*), both of which exude a milky liquid. This liquid, by no means as benign as milk, is sometimes heated, or powdered, then mixed with water and other ingredients to attain a certain consistency and concentration before it is used to coat the dart-heads.

The Dayak or Punan hunter carries two quivers on his back, one for the dart shafts, about 20 centimetres long, and the other for his poisoned dart-heads, each of them carefully rendered poisonous to differing degrees, according to his specific needs. Birds do not require any poison at all, just a plain dart to stun them. As the dart pierces the prey, the shaft drops off, leaving the head embedded. There can be no escape for the hapless victim. Death comes within minutes, the paralysing poison acting directly on the central nervous system. Before eating his quarry, the hunter will carefully cut out the flesh surrounding the dart wound.

It is unlikely that human beings will suffer an *ipoh* dart-wound nowadays, but should this occur, a traditional treatment is to make the wound bleed profusely (having extracted the dart-head without breaking it off), and smear on the pungent fermented shrimp-and-chilli paste essential to the Malay kitchen, or else the juice of a local lime.

Derris is commonly in use as an insecticidal powder, on sale at Western supermarkets. To Man, it is only cumulatively dangerous. If ingested in quantity or high concentration, however, it will stimulate salivation, produce a numb feeling around the mouth and tongue, and ultimately affect speech. Of this effect, colonial naturalist Charles Hose wryly remarked in 1929: 'In view of the increasing spread of democracy, it is possible that up to now its value has been underestimated.'

Locomotives of the Skies

Nobody who has been in the Kalimantan rain forest could forget the sight and sound of a flight of hornbills majestically flapping their way over the tree canopy. These magnificent birds, burdened with massive lumpy or casqued bills, thrash the air with their wings making a booming sound as they whoosh overhead like some heavenly locomotive.

At rest high in the forest trees, a hornbill's spectacular voice echoes through the stillness as a resonant honk, which in the helmeted hornbill (*Rhinoplax vigil*) builds itself up to a hysterical crescendo, peaking in a climactic cackle. This is one of the most thrilling sounds of the forest—or one of the most disturbing if you are not used to it. Mothers-in-law particularly have cause to worry, if Kalimantan folklore is to believed, for the helmeted hornbill's call is so maniacal that it has been locally nicknamed 'the chop-down-your-mother-in-law bird'—as if it were the triumphant laughter of a young Dayak man excitedly sawing through the stilt-legs of his mother-in-law's village hut.

Hornbills belong to the bird family called Bucerotidae. Despite their superficially similar appearance, they are not related to the South American toucan. Eight of Indonesia's 14 hornbill species (all of which are protected) are found in Kalimantan, including the exotically named wrinkled hornbill (*Rhyticeros corrugatus*), wreathed hornbill (*Rhyticeros undulatus*), helmeted hornbill, and rhinoceros hornbill (*Buceros rhinoceros*). There are also the black (*Anthracoceros malayanus*) and the pied (*Anthracoceros albirostris*) hornbills, besides the bushy-crested (*Anorrhinus galeritus*) and white-crowned (*Berenicornis comatus*) species. All of these sport a generally black-and-white plumage. The wreathed hornbill is the strongest flyer, and has a call resembling a dog's bark, compared with the bushy-crested hornbill's voice, more reminiscent of a yelping puppy, while the call of the pied bird is

The large upturned casque, huge size, black tail stripe, and resonant honking call all tell the observer that this is a rhinoceros hornbill (*Buceros rhinoceros*), among the most impressive of Kalimantan's hornbills. The Dayak peoples associate this bird with their bird-god, Singalong Burong.

The bushy-crested hornbill (*Anorrhinus galeritus*) is unassuming, compared with some other hornbill species, but is nevertheless distinctive for its droopy crest. Its call is a puppy-like squeal.

goose-like, that of the black like an angry pig's squealing. To promote local pride in the birds, the Indonesian government has adopted several hornbill species as official symbols for specific provinces of the republic.

If mere Westerners find the hornbill an awesome bird, how much more the Dayaks—particularly the Kenyah, the Kayan, and the Ngadju—who revere it on a spiritual level, believing the hornbill is a symbol of death followed by resurrection. The reason for this belief becomes obvious with closer inspection of the bird's breeding habits: the female hornbill has a strange habit of walling herself into naturally formed tree-trunk cavities by sealing the hole with mud and twigs, and even her own droppings, with more than a little help from her male partner. While she is entombed in the tree, incubating her eggs, the male returns periodically to feed her through a tiny slit in the sealed nest cavity. Remarkably, if her 'husband' fails her, or dies, another male, or several males, will appear to carry on the feeding. At last, with her eggs safely hatched, the female breaks down the wall and emerges, reborn as it were, but fat, filthy, and very, very stiff. The birds then re-plaster the wall and the chicks stay inside, with both parents feeding them through the hole, until they are old enough to fly. It is no easy matter to find precisely the right tree-hole for raising a hornbill family, so not surprisingly, hornbill couples return to the same hole for many years in succession.

Both the mother hornbill and her young neatly evacuate their droppings outside the tree-hole by pointing their rears outside and defecating in a slightly explosive 'jet-propelled' manner. The seeds that fall to the ground during the feeding and defecation sessions soon germinate, and forest peoples say they can tell what age the young birds are by assessing the growth status of these plants at the foot of the nest-hole tree.

Hornbills are more generally revered, too, as representatives of the Dayaks' bird-god, Singalong Burong, and their feathers feature prominently in ceremonial robes and head-dresses, particularly those used during ceremonial dances, as well as on carved ornamentation of longhouse huts. The rhinoceros species is a central figure in the Ibans' Gawai Kenyalang annual thanksgiving ritual, *kenyalang* being the bird's name in the Iban language. A carved representation of the bird is the centre-piece of this ceremony. The birds are also closely observed as omen-augurs. For example, the flight direction of a flock of singing helmeted hornbills may indicate an imminent rainstorm. In one Punan story, the hornbill is said to guard the tree-trunk that bridges the chasm between life and death. The hornbill sentinel will help a successful head-hunter over the bridge, but will ignore any poor soul who has never taken a head, leaving him to tumble into the abyss below, there to become dinner for a giant fish.

Despite their mystic role in Dayak culture, and despite the young birds' looking like 'hideous naked squabs with protuberant abdomens and loose, wrinkled skin', to quote naturalist Robert Shelford writing in 1916, the hornbill's chicks are considered food by the Dayaks, and eaten raw.

Although spectacular in flight, hornbills usually present a rather unimpressive, ungainly hopping motion once they settle on tree branches. Their disproportionately large and seemingly heavy bills are, in fact, hollow or filled with spongy cellular tissue. The exception is the helmeted species, which has a solid bill featuring a massive casque protruding above the main bill, from which the bird derives its name. It is this reddish-hued, solid bill which traditionally has been much sought after for 'hornbill ivory' or *ho-ting*, valued at least as highly as, if not more than, elephant tusks, and higher even than jade, by Dayak and Chinese ivory-carvers. Dayak peoples make delicately looped ear-rings out of this 'ivory'.

Not enough is known of these great birds' life cycles and habits. While the hornbills seem to live largely on fruit, especially figs, they also eat animal prey from time to time, making them more or less omnivorous. One thing is for sure, though—hornbills need very tall trees, with natural holes in them, to survive, and these most often prove to be dipterocarp or hardwood trees.

While hornbills make good pets, being easily tamed to the point where they will follow their owners about like dogs, there can be no sight as rewarding as the free flight of hornbills in their natural forest habitat.

The sound in the sky is unmistakable when a hornbill flies overhead, an awesome sight. Here is the great hornbill (*Buceros bicornis*) in flight—this species is found in Sumatra, but not in Kalimantan.

7 Sulawesi: Wallace's Think-tank

THE island of Sulawesi in itself resembles a living creature of some kind, like some microscopic amoeba adorned with flagellating antennae. Less an island than a spider-like assemblage of four interconnected peninsulas, the product of the collision of ancient continental shelves, this 195 000-square-kilometre blob offers quite extraordinary natural history interest. A mysterious land still bubbling with volcanic activity, evident in hot mud and hot water springs and sulphurous blow-holes, Sulawesi offers the thrill of the unexplored and unknown, even rumours of lost tribes deep in the interior.

This is the first time that this book has crossed the fabled, and imaginary, 'Wallace's Line' which falls between Bali and Lombok, and between Borneo and Sulawesi, defining the beginning of a transitional zone between the Oriental or Asian-style natural history of the Greater Sunda islands of Sumatra, Borneo, Java, and Bali, and the Australian character of life-forms in Sulawesi, the Moluccas, the Lesser Sundas (and the Philippines, a separate nation), and Irian Jaya beyond. Alfred Russel Wallace, travelling through the region in the 1850s collecting animal specimens for museums, independently came to the same conclusions as the more famous 'Father of Evolution', Charles Darwin.

Wallace's is not a rigidly fixed line. Scientists have enough exceptions, doubts, and counter-proposals to keep the arguments going interminably, but for our broad purposes here, the Wallace's Line concept performs a useful function. The animals, less so the plants, found on either side of this line differ in character. Broadly speaking, on the Sulawesi side, they move closer to Australian species, with possums (*Dactylopsila* spp.) and eucalyptus trees among the more obvious links. The only eucalypt or gum tree found in rain forest anywhere is the *Eucalyptus deglupta* commonly seen along rivers in Sulawesi, easily spotted by its impressive size and peeling reddish bark. As a result of Wallace's work, the unit formed by Sulawesi, the Moluccas, and the islands of the Lesser Sundas, the latter known as Nusa Tenggara in Indonesia, is referred to by biologists as 'Wallacea'. The whole region acts as a transition zone between Asian and Australian life-forms before you get to the more decisively Australian fauna of Irian Jaya in easternmost Indonesia, and Australia beyond.

The island of Sulawesi differs from Java, Sumatra, and Borneo (Kalimantan) in that it was not connected by land with the Asian continent during the last Ice Age. Human culture in Sulawesi, therefore, is much younger than in Java, for example, dating back only about 35,000 years. As in Australia, stencilled human hands on cave walls in Sulawesi send a moving message across tens of thousands of years to communicate with us today. Certain areas of Central Sulawesi (Lore Lindu National Park, for

◁
The day dawns over Soroako forest in South Sulawesi. This attractive area is now witnessing development, mostly associated with nickel-processing. It hosts one of the deepest lakes in Sulawesi, Lake Matana, 600 metres deep.

135

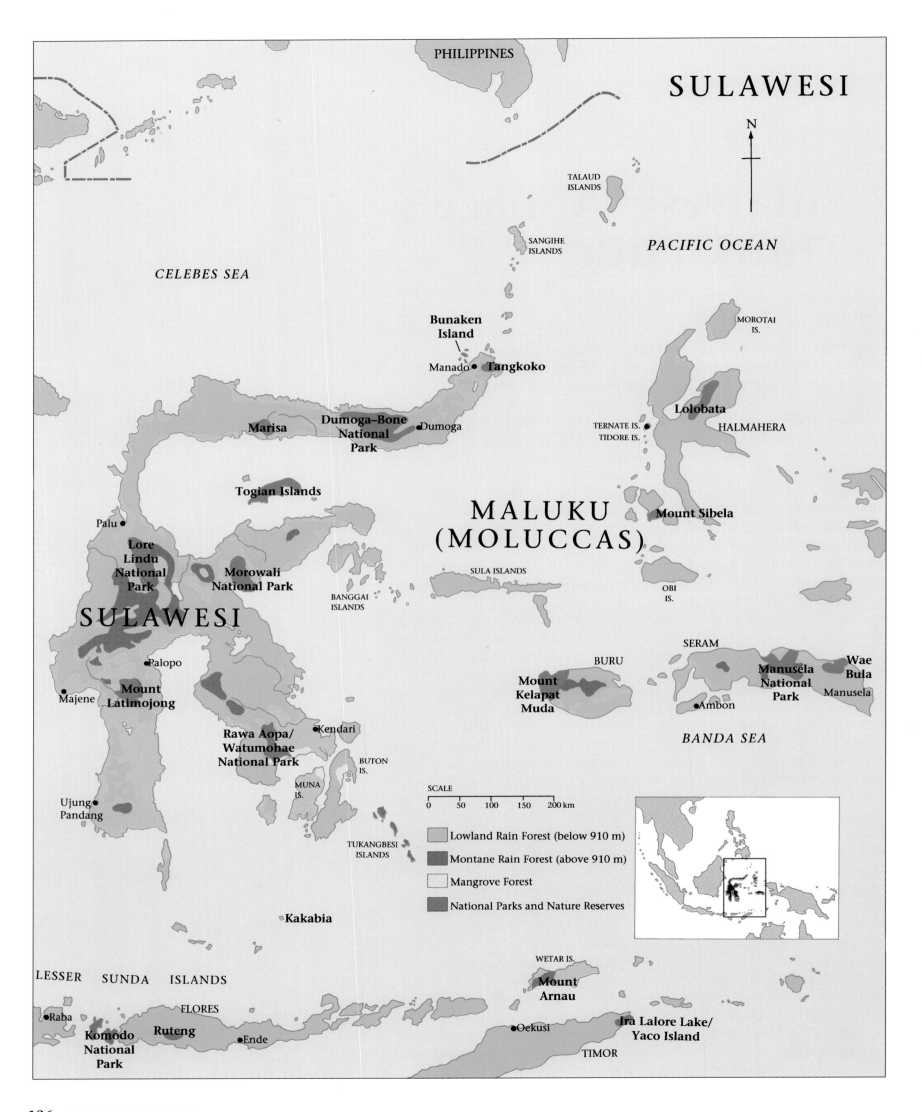

SULAWESI

PHILIPPINES

N

PACIFIC OCEAN

CELEBES SEA

TALAUD
ISLANDS

SANGIHE
ISLANDS

MOROTAI
IS.

Bunaken
Island

Manado ● Tangkoko

Marisa

Dumoga–Bone
National
Park

Dumoga

Lolobata

TERNATE IS. ●

HALMAHERA

TIDORE IS.

Togian Islands

MALUKU
(MOLUCCAS)

Mount Sibela

Palu ●

Lore
Lindu
National
Park

Morowali
National Park

SULA ISLANDS

OBI
IS.

BANGGAI
ISLANDS

SULAWESI

SERAM

Palopo ●

BURU

Manusela
National
Park

Wae
Bula

Mount
Kelapat
Muda

Majene ●

Mount
Latimojong

Manusela

● Ambon

Rawa Aopa/
Watumohae
National Park

Kendari ●

BANDA SEA

BUTON
IS.

Ujung ●
Pandang

MUNA
IS.

SCALE

0   50   100   150   200 km

Lowland Rain Forest (below 910 m)

Montane Rain Forest (above 910 m)

Mangrove Forest

National Parks and Nature Reserves

TUKANGBESI
ISLANDS

Kakabia

WETAR IS.

LESSER   SUNDA   ISLANDS

Mount
Arnau

Raba ●

FLORES

● Oekusi

Ira Lalore Lake/
Yaco Island

Komodo
National
Park

Ruteng

● Ende

TIMOR

example) are scattered with strange megalithic statues and stone vats, largely unexplained. Other evidence seems to link Sulawesi's early peoples with the Bronze Age cultures of Vietnam and Laos.

Again, unlike Java or Sumatra, Sulawesi historically had little contact with the Hindu culture. The island is still alive with ancient animistic ritual and the drummings of magic-wielding shamans. Nothing ties people more closely to wild Nature than animism, endowing as it does every rock, every tree, and every rainbow with its own life-force and spiritual power.

Best known of the Sulawesi peoples are the Bugis, warlike explorer-sailors and traders who have dispersed all over Indonesia and to Malaya, the highlander Toraja with their complex funeral rites and strange cave-burial customs, and the Bajau 'sea-gypsies'. Bugis praos or *pinisi,* skilfully crafted wooden sailing vessels, were—and still are—a common sight off the northern coast of Australia, collecting trepang or sea-cucumber (Holothuroidea)—also known popularly as sea-slugs, which in fact are biologically different animals—for the hungry Chinese markets of South-East Asia, or off Singapore, trading rare birds and other wildlife, or vibrant Sulawesi silks, for coveted consumer goods.

While mountainous Sulawesi—still known to many by its former name, Celebes—is handsomely endowed with primary tropical rain forest, its total combination of several forest types covers more than 50 per cent of its surface. It also houses Indonesia's most arid region, hospitable to the introduced American prickly pear cactus. Rich in oil, natural gas, and ore-bearing rock, not to mention gold, Sulawesi is an eldorado as yet only partially exploited. So it is too for the naturalist. Of the island's 127 native mammals, 79 species are endemic, many of them described as 'primitive'. These include notable oddities like the babirusa or deer-pig (*Babyrousa babyrussa*) and the lowland anoa or miniature buffalo (*Bubalus depressicornis*), besides 7 different kinds of crested black (*Macaca nigra*) or brown (*Macaca brunnescens*) macaques, the tiny 10-centimetre-long spectral tarsier (*Tarsius spectrum*) with its swivelling head, and 2 ponderous cuscus,

Endemic to Sulawesi, the crested black macaque (*Macaca nigra*) is tailless, like all the seven species of Sulawesi macaques. Seen here at Dumoga–Bone National Park, it patrols the forest canopy in unusually large troops.

Unbearably cute, these spectral tarsiers (*Tarsius spectrum*) peer from their bedroom, a tree-hole in the Tangkoko Reserve, North Sulawesi. This tiny primate is known for its family singing ritual before retiring at dawn to sleep during the day.

both marsupials—the day-feeding, leaf-eating bear cuscus (*Phalanger ursinus*), about 1 metre long, and the much smaller nocturnal, fruit-and-insect eating dwarf cuscus (*Phalanger celebensis*), both possum relatives. Many of the 68 bats found on the island are also endemic. One well-known one is the 'Harlequin Bat' or Wallace's fruit bat (*Styloctenium wallacei*), with its attractively brown-and-white striped face and back.

In the mossy mountain forests are many species of rat (Sulawesi has more than 40 altogether), including about 9 types of shrew-rat (Muridae), named for their long shrew-like noses, the 10-centimetre-long red tree mouse, Sulawesi pygmy tree mouse (*Haeromys minahassae*), and other small mammals such as ground squirrels (Sciuridae). The shrew-rats illustrate the principle of ecological niches very well, since each species thoughtfully eats something entirely different from the other (ranging from worms to fruit), frequents a different part of the forest, or forages at a different time (be it night or day). Thus, these rather similar animals take care not to compete or interfere with each other's survival. Also living high up is the Sulawesi (or giant) civet (*Macrogalidia musschenbroeckii*), an elusive animal with impressive climbing, or rather, descending, abilities; it can descend vertical slopes head first. Sulawesi's only native carnivore, this civet is very rare indeed.

Sulawesi lays claim to having recorded the world's longest snake, a 9.97-metre-long reticulated python (*Python reticulatus*), among its 64 snake species, 15 of them endemic. The fearsome estuarine crocodile (*Crocodylus porosus*) is also omnipresent.

Sulawesi produces alarmingly long reticulated pythons (*Python reticulatus*). The specimen here is about 6 metres long.

Australian-type species such as this bear cuscus (*Phalanger ursinus*) signal that Sulawesi is a transitional zone between typical Asian and typical Australian fauna. This phalanger moves around slowly through the treetops in small groups, using its prehensile tail as an extra limb while it nibbles leaves and fruit.

The knobbed hornbill (*Rhyticeros cassidix*), also known as the red-knobbed hornbill, is one of two hornbill species endemic to Sulawesi, and the larger of the two. As with other Indonesian hornbills, its feathers are often used in ceremonial dance costumes on Sulawesi.

The beauty and brilliancy of this insect are indescribable, and none but a naturalist can understand the intense excitement I experienced when I at length captured it. On taking it out of my net and opening the glorious wings, my heart began to beat violently, the blood rushed to my head, and I felt much more like fainting than I have done when in apprehension of immediate death. I had a headache the rest of the day, so great was the excitement....

(Alfred Russel Wallace on viewing a birdwing butterfly in the Moluccas, 1858, in *The Malay Archipelago*, 1869)

▷
Lontar palms (*Borassus flabellifer*) at Dumoga–Bone National Park. This park is well stocked with all typical Sulawesi-endemic flora and fauna.

There are also 96 species of bird endemic to the Sulawesi region, of a total inventory of 385. These include 7 mynas, 10 owls, and 5 starlings. The most celebrated Sulawesi endemic is the maleo (*Macrocephalon maleo*) with its enormous eggs. But equally striking are the knobbed (*Rhyticeros cassidix*) and the Sulawesi (*Penelopides exarhatus*) hornbills, the helmeted myna (*Basilornis galeatus*), the finch-billed myna (*Scissirostrum dubium*), and the dark green purple-bearded bee-eater (*Meropogon forsteni*).

Every bit as attractive as the birds of Sulawesi are the island's 450 butterflies, whose beauty summoned ecstatic prose from Wallace during his nineteenth-century travels. You can still observe today one of the sights that so entranced him: clouds of brightly hued butterflies—orange, yellow, white, blue, and green—settled near small forest pools on a hot afternoon, which, once disturbed, rise into the air by the hundreds like the rainbow spheres of a child's bubble-blowing pipe. Most notable are 38 species of large swallowtail, 11 of them endemic to the island; the handsome black-and-white Palu swallowtail (*Atrophaneura palu*) is a good example of these. Another rare butterfly is the Tambusisi wood nymph (*Idea tambusisiana*) from Morowali National Park. The entomological expedition, 'Project Wallace', organized by the Royal Entomological Society of London in 1985, did much to advance studies of this little documented facet of Sulawesi's wildlife.

The rarest of all Sulawesi's creatures must surely be those which are super-endemic, in that they are found only on small islands off the northern tip of Sulawesi, and not on the Sulawesi mainland itself. Examples of these highly vulnerable animals are the red-and-blue lory birds (*Eos histrio*) from the Talaud Islands, and the Talaud black birdwing butterfly (*Troides dohertyi*) on Talaud Island itself.

Unhappily, just about everything in this wildlife cornucopia, from forest rats to cuscus and fruit bats, seems to end up in Sulawesi markets, destined for the table.

The botany of Sulawesi has been little documented and so far only 7 endemic plant species are known. Moreover, the island's lowland forests feature only 6 species of the tall dipterocarp trees that are so plentiful in Borneo, Sumatra, and Java—with 267 species in Borneo, and over 100 in both Java and Sumatra. In the lowland forests, one especially valuable tree species is typical of Sulawesi: ebony (*Diospyros* sp.). Other commercially valuable timbers are present, including the historically introduced teak (*Tectona grandis*), and agathis (*Agathis* sp.), besides non-timber forest produce such as rattan. As in other parts of Indonesia, a very popular

Sulawesi's store of Lepidoptera is richly stocked. This swallowtail (Papilionidae) is just one of 450 butterflies found on the island.

Phalaenopsis cornucervi, one of many orchid species found on Sulawesi.

plantation crop is the clove (*Eugenia aromatica*), followed closely by the cashew (*Anacardium occidentale*). In Central Sulawesi, peoples of the interior without knowledge of weaving produce very fine bark cloth fabrics from trees such as the breadfruit (*Artocarpus altilis*) and wild fig (*Ficus* spp.).

Palms, especially the round-leaved fan palm (*Livistona rotundifolia*), and rattans abound in the Sulawesi forest. Typical, although also found in the Moluccas, is the tall *Pigafetta filaris* palm, specially adapted to survival in disturbed areas where the light is brighter than in most rain forests. This splendid palm, a prolific fruiter, sports shiny golden spines at the base of its leaves and a dark green trunk ringed with grey marks where the older leaves have fallen off.

The major wilderness areas of interest in Sulawesi are Dumoga–Bone National Park in North Sulawesi; Lore Lindu National Park in Central Sulawesi, where Indonesia's lowest rainfall is recorded; Morowali Nature Reserve in the east with its five roaring rivers, remarkable for a highly representative sample of Sulawesi's birds and rich mammalian wildlife; and the Tangkoko–Batuangus–Dua Saudara Reserve (named after three volcanic mountains in the reserve), east of Manado at the island's northernmost tip.

This is the flower of a shrub (*Clerodendrum* sp.) found at the lower levels of the forest, which is attractive enough to be sought out for cultivation as a pot-plant. The plant is interesting for being a hermaphrodite, carrying both male and female reproductive features.

Round-leaved fan palm (*Livistona rotundifolia*) forests are typical in lowland Sulawesi. This palm has a spiny stem and large leaves which may grow to 130 centimetres in diameter.

A glistening cicada (Cicadidae) emerges from its chrysalis at Dumoga–Bone National Park. Wingless cicada nymphs live in the soil but tunnel up to the surface and then climb tree-trunks for this final moult, becoming the winged adult insect. The incessant whining of male cicadas is one of the most familiar sounds of the rain forest.

A stick insect (Phasmatidae), its wings unusually displayed, seen in Dumoga–Bone National Park. Insects rank among the least studied wildlife of the rain forest. More than half the currently known species of the world are insects, and it is believed that thousands, if not millions more, remain unknown as yet.

The Sulawesi wilderness is tough trekking country, with one-fifth of the island above 1000 metres. Dumoga–Bone, for example, consists mainly of steep ridges. Its highest point, the summit of Mount Ganbuta, is almost 2000 metres. Fan palms, rattans, and eucalyptus trees, as well as pandanus, and moss forest at the higher altitudes, are among the special botanical features of this 3000-square-kilometre park, a major water-catchment area whose integrity is protected partly with World Bank assistance.

A large number of Sulawesi's endemic birds are found in Dumoga–Bone, from the spot-tailed goshawk (*Accipiter trinotatus*) and the Sulawesi scops owl (*Otus manadensis*) to the bay coucal (*Centropus celebensis*), as well as mammalian rarities such as the anoa and babirusa, and three of Sulawesi's endemic macaques. The frequently heard soprano songs of the spectral tarsier are often mistaken for insect whines. This was the base area for the 'Project Wallace' expedition in 1985, so the richness of insect, and particularly butterfly, life can be imagined.

Lore Lindu covers 2310 square kilometres of mostly montane forest, rich in wildlife. Here, the montane forests are dominated by higher-altitude species such as oak and conifer trees. The area is notorious for being the only place in Indonesia where the parasite which causes the dangerous disease schistosomiasis is found.

The spectacular 90-square-kilometre coastal Tangkoko area is home to the increasingly rare maleo bird, the bear cuscus, the tarsier, macaques, and wild pigs, besides many birds, including the lilac-cheeked kingfisher (*Cittura cyanotis*) and the Sula scrubfowl (*Megapodius bernsteinii*), a relative of the maleo. Crested macaques are particularly numerous here, feasting off the reserve's many and bountiful fig-trees. Tarsiers, too, are plentiful.

Kathy MacKinnon has likened Sulawesi to a biological laboratory where experiments in evolution have taken place. Isolated and strange, Sulawesi has given birth to unique fauna and a very rich avifauna.

A Tree Worth Preserving

Very few conifers are adapted to living in the rain forest, but the agathis tree (*Agathis* spp.) found in Sulawesi, and also in Kalimantan, is one. The agathis is a 'gymnosperm' or a primitive plant which seeds without flowering, and it is especially tolerant of poor soils lacking nutrition, and of steep slopes.

The tall, straight agathis, with its greyish leaves and round-scaled, flaky-barked trunk, is both beautiful and valuable, yielding a very high density of workable pale and fine-grained timber per hectare. It can grow to 50 metres in height, with a girth of 1.5 metres. Attempts to grow the agathis as a plantation tree have been reasonably successful, even outside Sulawesi, on Java. Different species of agathis are widely distributed, from the Malaysian peninsula to Fiji, Queensland in Australia, and New Zealand (where a close relative is known as the kauri tree), but it does not naturally occur in the wild on Java, nor in the Lesser Sundas, nor in Irian Jaya, except as a planted alien. Several species of agathis are tapped for *damar*, the tree's resin, which is also known as Manila copal and is used for making varnish, lacquer, and linoleum, as well as incense.

Agathis stands have long been decimated in New Zealand, and are disappearing in Kalimantan, too. Sulawesi could be one of the tree's last retreats. Conservation efforts made in parks such as Sulawesi's Lore Lindu offer hope for its survival.

An agathis tree (*Agathis* sp.) stands tall, shrouded in luxuriant lichen. A conifer, the agathis is a primitive plant with considerable commercial value.

Kingdoms of the Sun

Before the coming of Islam, the mighty kings of Sulawesi—southern Sulawesi, in particular—worshipped the forces of Nature in the shape of the sun and the moon, orienting their graves east–west. Courtly rituals must have been grotesque yet resplendent, for the ruler was seen as a living fertility symbol and was assisted in the performance of various arcane ceremonies by transvestite priests. Much of this culture remains buried in the murky past, for the Bugis–Makassar script was invented only in about the year AD 1400.

The oldest and most powerful of the Sulawesi kingdoms was Luwu, founded in the first millennium AD, when—as legend would have it—the gods first descended to earth. The kings claimed their descent from these divine ancestors. In their world, status and pedigree were all important; 'marrying down' was a capital offence. The divine founder of Luwu was Simpurusia, or 'lion-man'. His wife was said to have risen from the seas, carrying a name which translates as 'she who trapped her lord in the snares of a net'.

By the fifteenth century, Luwu controlled the east and south coasts, and the west coast as far as Makassar (Ujung Pandang), source of the famous Makassar oil derived from a local tree, which Western men once loved to lard upon their hair. English housewives used to place protective lacy napkins called 'macassars' over the head-rests of their armchairs to ward off the ravages of this hair oil. Luwu grew rich from trading in the natural resources of the Sulawesi soils and forests, including gold and tree resins. However, in the sixteenth century, the might of Luwu was at last humbled by the kingdom of Bone, and Luwu faded away.

The Dinosaur Bird

The endangered maleo (*Macrocephalon maleo*) is a weird and wonderful bird classified as a 'megapode', meaning 'giant foot'. It is a chicken or bush turkey in that it incubates its eggs at ground level. The maleo has relatives in Australia and Polynesia, but is found only on Sulawesi in Indonesia.

About as big as the domestic hen, the maleo has a perkily erect black tail, black back, pinkish-white belly, and a greenish beak with a red base. Its strangest feature is its distinctively black-knobbed head. 'The appearance of the bird when walking on the beach is very handsome. The glossy black and rosy white of the plumage, the helmeted head and elevated tail, like that of the common fowl, give a striking character, which their stately and somewhat sedate walk renders still more remarkable,' commented Wallace.

Population estimates dating from the late 1970s suggest there may be between 5,000 and 10,000 maleos on Sulawesi, but this is not confirmed. Unlike its Australian megapode relatives, the maleo does not construct

The maleo (*Macrocephalon maleo*), a Sulawesi endemic, has a knobbed head which some zoologists have theorized might be a sort of sun-helmet to protect the usually forest-dwelling bird from the sun while it digs its beach-sand egg-incubation holes out in the open.

mounds of rotting vegetation to warm its eggs, but instead digs pits in warm ash or sand. Another smaller megapode, also found on Sulawesi, the Sula scrubfowl (*Megapodius bernsteinii*), prefers to lay its eggs in burrows between the decaying roots of trees.

Each pink to red egg, about 11 centimetres long, and weighing more than 250 grams, is twice as big and five times as heavy as the common hen's egg. These large eggs supply a proportionately large yolk-store for the developing young to feed off, but they also make good, protein-rich eating for human beings. This is a temptation for many who risk breaking the law by poaching them, if not for the nutrition, then for the US$0.80 a single egg can fetch on the market (about five times the price paid for chicken eggs). Maleo egg harvesting was once jealously controlled by pirate kings. Ironically, such undemocratic procedures were more conducive to maleo conservation than the blatant collection which sometimes takes place today.

Courting maleo couples make quite a sight—and sound—as they cluck like a pair of duck-turkey hybrids while scratching out a deep, 1-metre pit in the

A maleo couple sets about the frantic labour of digging a hole for their eggs. They share the work, which may take them about an hour. After the female has laid a single egg in the hole, the two birds will then fill it in, and set about digging yet another hole nearby, just as a decoy to confuse predators, such as the monitor lizard (*Varanus salvator*). In 10 days' time, they will do it all over again. Meanwhile, the hot sun beating on the sand acts as a natural egg-incubator.

ground with their slightly webbed toes. This task probably helps to explain why they have such huge feet. They may work at this for about an hour, and it does seem as if their oddly helmeted heads are designed to protect their brains against the blazing sun while they do this. Their preferred habitat is shady forest. Another possibility is that they use the knob on the skull to gauge the temperature of the ground before burying their eggs. With a 7–10 day interval between each laying, and about 8–12 layings a year, each egg gets its own individual pit. Some cleverly placed decoy pits are deliberately designed to fool egg-eating predators such as monitor lizards (*Varanus salvator*). Unfortunately, however, this stratagem is not enough to fool humans.

Thus protected, but with their parents long departed from the scene, the young hatch from the eggs after three months' incubation in warm volcanic soils or sun-bathed black beach sand, sometimes heated by nearby volcanic vents to the requisite 32–38 °C. The chicks burst forth fully plumed, air-borne, and independent, due probably to their intake of the high-protein yolk, after about two days of labour, and immediately run for lowland forest cover, their natural habitat. The fact that their parents take no part in their incubation, unlike domestic chickens, and that the eggs are buried, is reptilian in character, and indeed, these birds are direct descendants of small dinosaurs.

You can easily spot a communal maleo breeding ground: it looks as though an army has just passed through on exercises, leaving bomb craters behind. In extreme cases, as many as 600 nest-pits may be on the same breeding ground. There may be 50–60 such grounds in all Sulawesi. Nobody quite understands how the adult maleo knows where to go when its time to breed comes, since the nesting area may be as far as 9 kilometres from its home range. Normally, the bird is found scratching for insects or fallen fruit on the forest floor. It does fly, although ponderously, and roosts on low-lying tree branches.

The maleo needs protection. The tactic adopted by Dutch ornithologist Dr Rene Dekker, working for the International Council for Bird Preservation (now Birdlife International) in Dumoga–Bone National Park, is to collect and rebury a designated number of maleo eggs inside a protective cage, to give maximum protection against both humans and other predators like pigs, dogs, and lizards. Properly managed, this yields an 80 per cent hatching record. Eggs laid outside protected or nature reserve areas can be freely harvested by the local people. Another potential source of maleo revenue is maleo tourism, conducted along the same lines as turtle-watching on the east coast of Peninsular Malaysia.

A sectioned maleo incubation-hole shows the young maleo, recently emerged from the egg. Incubation takes about 80 days. When the young bird hatches, it must struggle up through the sand to the surface in order to reach its natural habitat, the forest.

It has been about three days' hard work coming up from under. The newly hatched maleo is already fully feathered and now bursts out of its underground incubator, to rush to the nearest forest refuge.

The Mini Buffalo

Though deer-like in appearance, with short, sharp horns, the anoa found only on Sulawesi, is the world's smallest buffalo. This sturdy dwarf buffalo does not herd but moves about the mountainous areas of Sulawesi singly, or in small groups. There are two distinct species of anoa: the smaller mountain species (*Bubalus quarlesi*), 75 centimetres at the shoulder, and the white-stockinged lowland species (*Bubalus depressicornis*), 1 metre tall, although the distinction between them may not seem very clear-cut to the layman or casual observer.

Anoa graze on grasses, ferns, the leaves of gingers, pandanus, certain trunk-fruiting figs, and even mosses, and search the seaside grasslands at night for water and salt; they love to wallow and need mineral salts to complete their diet. Except for these forays out of the forest on to the beach, anoa differ from other wild South-East Asian cattle, which are happy at the edge of the forest or in secondary growth, in that they need deep, undisturbed forest for their natural habitat. This lack of adaptability makes the anoa's survival very fragile indeed.

Small though the anoa may be, it can hold its own in any confrontation with human beings, for its horns can deal a fatal blow if the animal is cornered into butting. Those who have attempted to keep the animal in captivity, often for its tasty meat, have mainly proved that the anoa is extremely difficult to tame.

The anoa (*Bubalus depressicornis*) is a dwarf buffalo, one of Sulawesi's several interesting endemic species. Its short sharp horns are capable of inflicting fatal injuries on humans. Although essentially a forest animal, the anoa searches constantly for sodium and may therefore venture into salt-water swamps and on to the seashore.

A Pig Like No Other

'He uses them to hang on branches when he goes to sleep.'

'They are to protect his eyes as he pushes through the undergrowth.'

'They are weapons of defence against predators.'

'They're simply attractive to female babirusa.'

The babirusa (*Babyrousa babyrussa*) or deer-pig's extraordinary curved tusks are actually upper incisors which grow through the male animal's muzzle, and curve backwards to form an arch in front of its eyes. The full length of these formidable tusks can reach 31 centimetres. From the lower jaw protrude two shorter, but well sharpened, tusks, regularly honed by rubbing against trees. As Wallace said in 1869, 'It is difficult to understand what can be the use of the extraordinary horn-like tusks.'

This bizarre creature, weighing about 100 kilograms, was first described to the West in 1658, and was apparently kept and bred in captivity by the kings and princes of Sulawesi, partly as a novel gift for important visitors. Pointing out that the male's tusks could hardly be meant to protect the animal from thorny shrubs, since the female would need them too, Wallace surmised that 'These tusks were once useful, and were then worn down as fast as they grew; but that changed conditions of life have rendered them unnecessary, and they now develop into a monstrous form, just as the incisors of the beaver or rabbit will go on growing, if the opposite teeth do not wear them away.'

Current scientific thinking favours the explanation that these tusks are actually used as weapons in competitive combat with other males: the curved upper tusk is hooked around an opponent's lower tusks, allowing the winning babirusa to continue stabbing at its enemy's exposed and undefended throat with its lower tusks. Such gladiatorial contests explain why many animals' tusks are broken or worn down and badly scratched.

The babirusa (*Babyrousa babyrussa*), seen here at Dumoga–Bone National Park, is neither a pig nor a deer, but a unique animal. Its extraordinary double pair of upward-curving tusks appear to function mainly as weapons of war among sparring males. The females do not have large tusks.

Babirusa are found on some other Indonesian islands: in the Moluccas and in the Togian Islands off the north of Sulawesi. Interestingly these animals have smaller tusks and a different way of fighting: butting rather than wrestling. The babirusa is said to be able to swim well, which might explain the animal's presence on Buru.

Usually hairless, the babirusa is a peculiar creature by any standards. That it has survived at all is partly due to the Islamic proscription against eating pork. However, in terms of what can be considered *halal*, or permissible to Muslims as food, the babirusa is a borderline case, since in many important ways, it is unlike a pig. For instance, the animal does not have a cloven foot, and it has a complex stomach which brings it closer to being a ruminant than a pig—correctly slaughtered ruminants are permitted food for Muslims. Conservationists understandably are not eager to point this out to Indonesian Muslims. Despite this religious protection, babirusa meat is still found in the markets occasionally, since quite a few people in Sulawesi are Christians.

The fact that the babirusa sow is a far less prolific breeder than her competitor, the wild boar, makes the babirusa's future even more uncertain. Equipped with only two teats for suckling, babirusa sows produce only one or two young at a time, and these grow slowly. This restrained breeding pattern is another very un-piglike characteristic. It may be a response to the lack of enemies in Sulawesi, the python being the only possible predator. Furthermore, unlike pigs, the largely nocturnal babirusa eats fallen fruit, including the potentially poisonous pangi (*Pangium edule*) and coconuts (*Cocos nucifera*), exclusively sprouting ones, and breaks open fallen tree-trunks to get at beetle larvae, rather than snuffle around in the soil for roots and worms. Wallace noted that the babirusa resembled the wart-hogs of Africa, but concluded, for all that, that 'the babirusa stands completely isolated, having no resemblance to the pigs of any other part of the world'.

A young babirusa snuffles at its mother's head. The young are born toothless. Only much later do the young male's canines suddenly grow upwards, piercing through the skin of the cheeks to curve and point towards the skull.

The babirusa eats fallen fruit such as coconuts and breaks open rotting fallen tree branches to look for beetle larvae. The animal is largely nocturnal and is said to be a good swimmer.

Where Spice Winds Blow

On the 999 remote and far-flung islands of the earthquake-prone Moluccas, the original Spice Islands whose aromatic treasures were once so coveted by Europeans, are little-known forests. This is a transitional zone, both for diverse Oriental and Australian species of fauna and flora, and also for the diverse peoples of the Moluccas, standing between Malay and Papuan racial types, with much mixing with the Dutch, Portuguese, and Javanese in between. There are still wandering bands of 'sea-gypsies', and in some areas, it is said that one-time head-hunters remain hostile to outsiders.

The Moluccas are now contained within the Indonesian province of Maluku. They include the better known islands of Halmahera (known as 'Gilolo' in Wallace's time), Seram (sometimes rendered 'Ceram')—these two being the largest—Ambon, Ternate, and Tidore, besides Buru, Wetar, the Banda Islands, and the Tanimbar, Babar, Aru, and Kai islands.

It was on the Moluccan island of Halmahera that Alfred Russel Wallace perfected his own ideas on the survival of the fittest, as he lay stricken with malaria, in 1858. His excitement prompted him to write an account of his ideas to Charles Darwin, who was astounded, and not a little disquieted, to receive such an exact reflection of his own developing ideas on evolution.

The two men should be, but unfortunately rarely are, thought of as joint partners in bringing about the philosophical and scientific revolution that the Theory of Evolution (usually attributed solely to Darwin) triggered in the West. Darwin himself took the wise advice of other, senior scientist friends and gave Wallace due credit. In Wallace's presence, he presented a jointly credited short paper to the Linnean Society in London in 1858, before publishing his own seminal work on evolution, *On the Origin of Species*, the following year.

However, the genesis of this important theory can hardly be attributed to anything particularly outstanding about the Moluccan fauna in terms of rain forest wildlife. True moist lowland forest is really only present on parts of Halmahera and Seram islands, which, however, are both large, each being about 20 000 square kilometres in area. There are in any case few land mammals of significance in the Moluccas. Among the endemic mammals are rats and bats, and marsupials such as a small flying possum. Others, such as cuscus, monkeys, pigs, and deer, were probably introduced by Man centuries ago. There is more interest in the life in and around the seas encircling the Moluccan islands, such as the Banda Sea, remarkable for huge

A swallowtail butterfly (*Papilio ulysses*); many swallowtails are beautiful to look at, but have a nasty taste for any predator that tries to eat them.

This is the world's largest Agamid lizard, the endangered sail-fin lizard (*Hydrosaurus amboinensis*), sometimes known as the Ambon dragon. The lizard is often found in or near water and has been known to run across the water's surface. Surprisingly, it is primarily a vegetarian, eating fruit and leaves.

breeding colonies of sea-birds, particularly around Manuk Island and Mount Api.

One interesting reptilian denizen of the Moluccas, also found in Sulawesi, New Guinea, and the Philippines, is the sail-fin lizard (*Hydrosaurus amboinensis*), or Ambon dragon of Ambon Island, which can grow to a length of 1 metre, most of this being the animal's tail. This lizard sports a sail-like crest running down its tail, and is unusual, for a lizard, in eating leaves. It can also run across the surface of water when threatened, with the aid of enlarged scales on its toes. It is related to similar lizards in Australia.

There are also some spectacular butterflies in the region, 25 of them endemic species, including 4 of the famous birdwings, such as the huge *Ornithoptera croesus*, with a 20-centimetre wing-span, glistening orange-brown on top, green and black below. Another intriguing denizen of the Moluccan forests is the world's largest bee (*Chalicodoma pluto*) at 4 centimetres long, which co-nests with termites, first reported by Wallace.

The avifauna of the Moluccas, however, is still more fascinating. Among the region's approximately 348 birds are 64 endemics. Conspicuous among the Moluccan birds, in terms of their vivid colouring, are 31 parrots, including the cockatoos. The region is home to the soft-pink salmon-crested cockatoo (*Cacatua moluccensis*) adept at opening up tough young green coconuts, the opulently coloured king bird of paradise (*Cicinnurus regius*), the rainbow lory (*Trichoglossus haematodus*), the eclectus parrot (*Eclectus roratus*), either green (male) or blue and red (female), and several small hanging parrots of the genus *Loriculus* which hang upside down like bats. The large two-wattled cassowary (*Casuarius casuarius*), more usually thought of in connection with New Guinea, is unexpectedly also found in the Moluccas' 1890-square-kilometre Manusela National Park, on Seram Island.

Interesting trees of the Moluccas, as well as Sulawesi, include the agathis (*Agathis* spp.), which yields commercially valuable resin, and the paperbark (*Melaleuca leucodendron*), which is a source of medicinal oils, and of course, the region's fabled spice trees, such as nutmeg (*Myristica fragrans*), clove (*Eugenia aromatica*), and cinnamon (*Cinnamomum* spp.). One native Moluccan species, the fast-growing and shady *Anthocephalus macrophyllus*, is an economically useful timber tree, while the kenari tree (*Canarium indicum*) is also typical of the Moluccas—kenari nuts are valuable as food (both to humans and to parrots), and as medicine.

The blue-streaked lory (*Eos reticulata*) is endemic to the Tanimbar Islands of the Moluccas, although it has been introduced to the neighbouring Kai Islands, and the nearby Damar Island. Lories probably play an important role in pollinating the trees and flowers on whose nectar they feed, crushing flowers to extract the sweet juices and collecting pollen with the brush-like tip to their tongue.

The eclectus parrot (*Eclectus roratus*) displays marked sexual dimorphism (differences between the sexes), in that the male is green, while the female is blue and red. For a long time, scientists mistakenly thought the two sexes were two different species of bird.

8 Glorious Irian Jaya

IRIAN JAYA is another world compared with the rest of Indonesia. In prehistoric times, it was probably part of Australia, integrated with Antarctica, India, South America, New Zealand, and New Caledonia, within the ancient super-continent of Gondwanaland. Only in geologically recent times, perhaps about 200 million years ago, did the island of New Guinea, on which the province of Irian Jaya sits, start to drift northwards towards Asia, together with Australia.

While Wallacea—Sulawesi, the Moluccas, and the Lesser Sundas—may be a transitional zone in terms of its fauna and flora, the animals of Irian Jaya place themselves firmly on the Australian side of the Wallace Line. For example, there are the flightless cassowary and 47 marsupials, or pouch-nursing mammals, such as the wallabies and kangaroos, as well as 2 types of spiny echidna, the short-beaked (*Tachyglossus aculeatus*) and the long-beaked (*Zaglossus bruijni*) species. Echidnas, together with the platypus (*Ornithorhynchus anatinus*), are the world's only egg-laying mammals, known as 'monotremes' to scientists.

◁
This Asmat from Irian Jaya is among the many tribal peoples who call the forest home—the forest feeds them (with sago in this picture).

The short-beaked echidna or spiny anteater (*Tachyglossus aculeatus*) is a monotreme, a mammal that lays eggs. The echidna's single egg is kept in a pouch on her belly, where it hatches and the young stays there, suckling on milk coming from slits in its mother's abdomen. When the young echidna's spines develop, the mother ejects it from the pouch.

The presence of wallabies in Irian Jaya underlines the increasingly Australian profile of the indigenous fauna. This is a Bruijn's pademelon or dusky wallaby (*Thylogale bruijni*), which can be found as high up as 4200 metres, in montane forests.

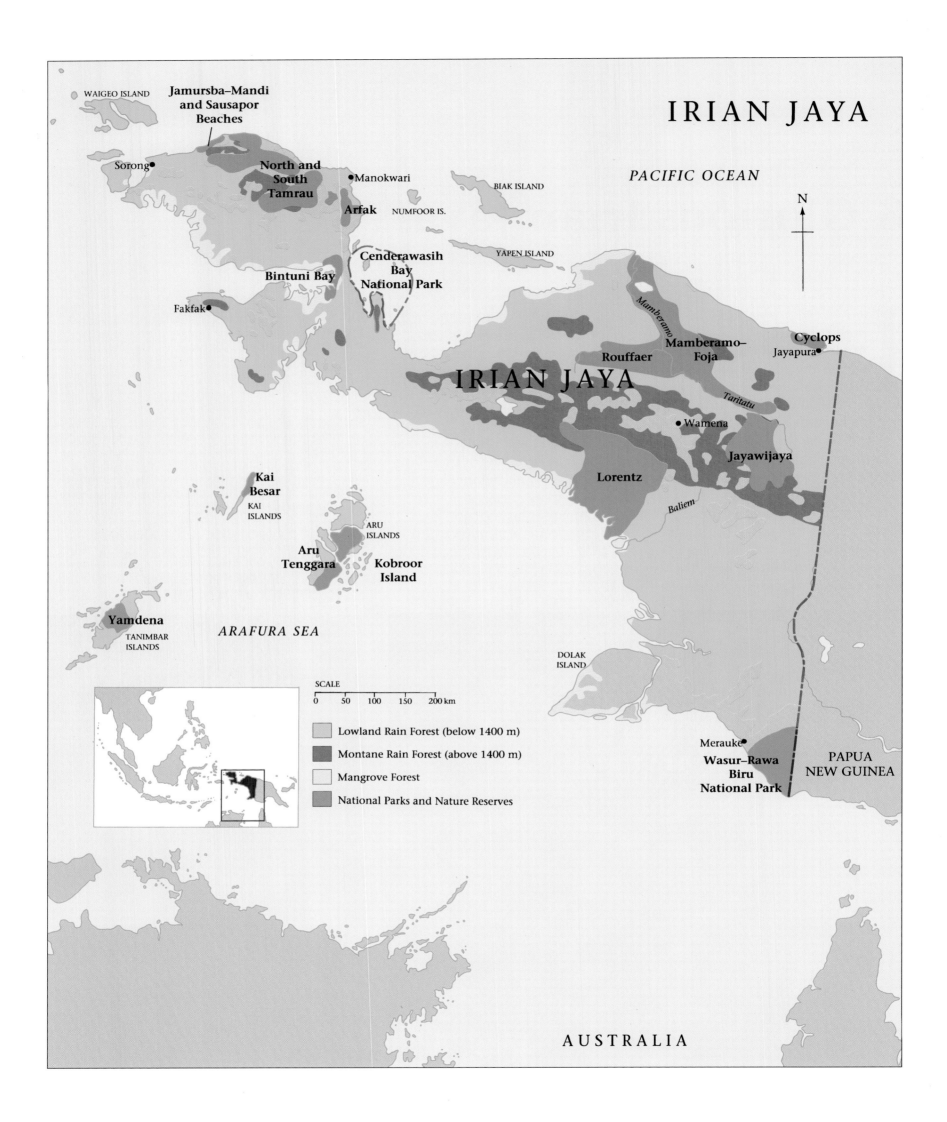

IRIAN JAYA

WAIGEO ISLAND

Jamursba–Mandi
and Sausapor
Beaches

PACIFIC OCEAN

Sorong•

North and
South
Tamrau

•Manokwari

BIAK ISLAND

Arfak

NUMFOOR IS.

N

YAPEN ISLAND

Cenderawasih
Bay
National Park

Bintuni Bay

Mamberamo

Cyclops

Fakfak•

Mamberamo–
Foja

Jayapura•

Rouffaer

IRIAN JAYA

Taritatu

Wamena•

Jayawijaya

Kai
Besar

Lorentz

Baliem

KAI
ISLANDS

ARU
ISLANDS

Aru
Tenggara

Kobroor
Island

Yamdena

ARAFURA SEA

TANIMBAR
ISLANDS

DOLAK
ISLAND

SCALE

0   50   100   150   200 km

Lowland Rain Forest (below 1400 m)

Montane Rain Forest (above 1400 m)

Mangrove Forest

National Parks and Nature Reserves

Merauke•

Wasur–Rawa
Biru
National Park

PAPUA
NEW GUINEA

AUSTRALIA

As the province's name translates, it is 'glorious' (Jaya) indeed. The country's zoological crown jewels are of course the 28 local species of bird of paradise, so exotically beautiful they hardly seem real. Freed of the constraints of major predators—apart from tribal, spear-wielding man, that is—the male birds of paradise have been able to evolve whatever fantastical, brilliantly colourful ornaments they please.

In terms of human culture, perhaps nowhere else in Indonesia, barring Sumatra's Siberut, is as close to the Stone Age as Irian Jaya. Plumed warriors with bones through their noses, and sporting only long penis-sheath gourds for clothes, still roam the land. Until as late as the 1970s, many tribesmen used only stone tools. Cannibalism lingers among isolated groups in some remote parts. The price of a human life—especially a female one—may be counted in pigs.

Most of the peoples of Irian Jaya—such as the Asmat, the Dani, and the Jale—are Papuan in origin, distant relatives of the similarly dark-skinned Australian aboriginals, with Austronesian, Malay-type intermixes more common in the north. Their roots in New Guinea go back as far as 40,000 years ago. In many cases, they remain remote, out of the reach of even modern communications systems. Parts of the interior are as yet unexplored and inadequately mapped; it was only in 1938 that Westerners discovered the great Baliem Valley in the central highlands, homeland of the Dani people.

Some groups practise quaint utopian 'cargo cults' which, though various in form, are based on the belief that the paradise represented by white men's goods, from machinery to medicines, can easily be obtained if the white man's 'rituals' are followed. In the 1930s, there were reports of some groups setting up tables and chairs and eating with knives and forks in the hope that this would bring them white prosperity.

Irian Jaya is a very special part of our planet that contains a vast wealth of natural treasures which stretch out over horizons of the most pristine environments that remain in the world. It houses the largest continuous tracts of undisturbed lowland rain forest in all of South-East Asia, and a biotic richness and diversity that is beyond compare. The province is one of the last great unknowns of the world … a challenge to understand, a challenge to explore, a challenge to develop, and a responsibility for all to safeguard its most important natural areas.

(Ronald G. Petocz, WWF representative in Irian Jaya during the early 1980s)

At the market in Wamena, the hub of the Baliem Valley, territory of the Dani tribe, pandanus fruit feature as a favourite food.

Because of intermittent tribal wars and tough terrain, encouraging isolation, each human group speaks a different language, so that New Guinea as a whole, with only about 0.05 per cent of the world's population and 0.15 per cent of the Earth's surface, harbours 15 per cent of the world's known languages—about 800 in all.

The land itself provokes childlike awe. For a start, it is huge, accounting for 22 per cent of Indonesia's entire land area. The island of New Guinea, of which Irian Jaya forms the western half—the independent nation of Papua New Guinea forms the eastern half—covers an area of 800 000 square kilometres, making it the world's second largest island after Greenland. Irian Jaya's share of this vastness is almost 420 000 square kilometres. Much of it is rich in primary resources, such as copper, gold, and silver. Huge as it is, it is home to only 1.7 million people, and large tracts of land are uninhabited, albeit claimed and used by various indigenous tribes.

This rugged, untamed terrain is an ecologist's playground, exhibiting everything from snow-capped peaks to lowland tropical rain forests, from dry eucalyptus savannah forests to peat swamps and extensive mangroves. About 82 per cent of the province is covered with undisturbed forest. The island of New Guinea as a whole boasts the world's second largest undisturbed rain forest area after the Amazon. This forest is particularly rich in plants of medicinal use, discovered and undiscovered. One such product is massoi (*Massoia aromatica*) bark, valued as a food flavouring by the Javanese, and by many Indonesians as a cure-all, particularly for fevers and also in the treatment of venereal disease.

Down the length of New Guinea runs a 2000-kilometre ribbon of jagged mountains, reaching heights over 4800 metres. These mountains are natural barriers to the spread of certain species, and account for many endemic animals and plants.

Spice trees such as nutmeg (*Myristica fragrans*) and clove (*Eugenia aromatica*), as well as precious timbers like ebony (*Diospyros* spp.), local forms of walnut (e.g. *Dracontomelon puberulum*), and sandalwood (*Santalum* spp.), are plentiful. One species of nutmeg has a 'partnering' relationship with ants which live in its hollow stems. In the south, the feathery casuarina tree, often mistaken for a fir, is common.

The matoa (*Pometia pinnata*), found only in the province, is an unusual native fruit tree, with small reddish-brown fruit hanging in grape-like bunches. One tree can produce 400 kilograms of fruit in a year. Easily

Irian Jaya rejoices in about 16,000 species of plant, with at least 124 genera endemic to the province. The flame of Irian (*Mucuna novaeguineensis*), a climber with spectacular hanging flowers, is one of the best known.

The sago worm, actually the larva of the Capricorn beetle, is an important source of protein for peoples such as the Asmat of Irian Jaya. The Asmat actually 'cultivate' the worm by deliberately creating the conditions it favours: boring holes for it in the rotting trunks of the sago palm.

peeled, the matoa has a sweet, refreshing taste, although it has a pungent smell. The untapped export market for such little-known foods can be guessed at.

The lowland dipterocarps present in Sumatra and Borneo, though fewer in number, can be found side by side with close relatives of the monkey-puzzle trees (*Araucaria cunninghamii*) which also grow in northern Australia's Queensland and in Chile, while higher up, northern hemisphere-type oaks (*Quercus* spp. and *Lithocarpus* spp.) jostle with Antarctic beeches (Fagaceae), also known from Australia, New Zealand, and South America. Most famous and most familiar of all is the flame of Irian (*Mucuna novaeguineensis*), a legume often cultivated in parks elsewhere for its drooping bunches of bright red flowers.

The Irianese themselves know their environment well and traditionally have used its natural resources, in almost every aspect of their lives: for food, magic, tools, weapons, construction materials, cloth, and medicines. One of the many household items they have created from the forest produce is the string-bag which tribal women suspend from their foreheads, particularly in the highlands. Knit from double-strength rolled bark fibres, these almost indestructible bags are used to carry everything from sweet potatoes to pigs and babies. They are delicately coloured by using clays, the yellowish and white outer skin of orchid tubers, and ferns for the red and brown shades. The bags reek of the wood-smoke and sweat which permeate the people's tiny huts.

A gourmet forest favourite with the Asmat people of the south, in particular, is the slug-sized larva of the Capricorn beetle, which lives in the sago palm (*Metroxylon sagu*), their other major source of food. The sago tree is indeed the staff of life for most Irianese, as rice is to most other Asians, and potatoes to Westerners.

It seems as if Nature had taken precautions that these her choicest treasures should not be made too common, and thus be undervalued. This northern coast of New Guinea is exposed to the full swell of the Pacific Ocean, and is rugged and harbourless. The country is all rocky and mountainous, covered everywhere with dense forests, offering in its swamps and precipices and serrated ridges an almost impassable barrier to the unknown interior....

(Alfred Russel Wallace, on his search for birds of paradise in 1860, in *The Malay Archipelago*, 1869)

Stilt-rooted screwpine forest (*Pandanus* sp.) is typical of the swampy areas in Irian Jaya. Some pandanus fruits are edible, but the leaves are particularly useful to Man, often dried for thatching or weaving materials.

This handsome tree frog (*Litoria caerulea*) is one of about 120 species of tree frog in Irian Jaya. It has specially adapted suction cups on its feet for climbing trees.

The Papuan tiger orchid (*Grammatophyllum papuanus*) is a huge orchid, the world's largest.

▷
A bevy of blue beauties—Victoria crowned pigeons (*Goura victoria*). These birds are quite used to humans and sometimes frequent sago-flour processing areas.

The sheer abundance and opulence of life-forms found in Irian Jaya is staggering. Only Borneo is comparable as a genetic treasure house. Irian Jaya's inventory includes 174 land mammals, 100 of them endemic to New Guinea; 643 species of bird, 269 of them endemic to New Guinea and possibly 40 or more endemic to Irian Jaya; about 91 species of frog; and at least 16,000 plant species, of which 90 per cent are New Guinea endemics. Irian Jaya also has most of the 9,000 flowering plant species found on New Guinea, including the majority of the 2,700 species of orchid, such as the world's largest (*Grammatophyllum papuanus*), exhibiting 3-metre orange sprays. Irian Jaya also accounts for a large proportion of New Guinea's 5,000 butterflies and moths, notably the huge birdwing butterflies centred in the Arfak Mountains, out of a total 100,000 insect species found throughout New Guinea. The level of endemism is high: about half of all Irian Jaya's plants and animals are found nowhere else but on the island of New Guinea.

The region is justly famous for its birds. While the birds of paradise naturally hog the limelight, there are many other species worthy of note. The Victoria crowned pigeon (*Goura victoria*), for example, is a delicate good-looking creature, with its soft blue-grey colouring and fragile tiara of crest-feathers, while 44 species of parrot illuminate the forests with their rainbow colours. There is also an extraordinary array of 23 kingfishers, more than half of Indonesia's total inventory of 45 species—including the quaint shovel-billed kookabura (*Clytoceyx rex*), which is partly nocturnal and uses its bill to dig up worms on the forest floor. Furthermore, the region boasts at least 58 species of Australian-type honeyeater, including the New Guinea-endemic orange-cheeked honeyeater (*Oreornis chrysogenys*),

besides another 5 species of honeyeater which are endemic to Irian Jaya.

Oddest of all is the bowerbird, of which there are 9 species in Irian Jaya. The female of the species is obviously something of a 'material girl', for her somewhat dull-looking suitor is obliged to compensate for his looks by courting her assiduously with all manner of precious and bright objects, from berries, flowers, and fruit to beetle skeletons, shells, pebbles, charcoal, and chewed grass, all artistically arrayed in an attractive nest 'apartment' or bower. Indeed, the Vogelkop bowerbird or bird's-head bowerbird (*Amblyornis inornatus*) goes as far as to build a full-scale 'hut' complete with front door, while others construct ceremonial 'malls' or fenced-in avenues.

There are other species which build 'maypole' columns of twigs 2 metres high or more to impress the female of their choice. However, once successfully mated, the male goes off, somewhat unchivalrously leaving his partner to get on with the business of raising their offspring alone.

The largest bird is the land-bound, sometimes vicious, ostrich-like cassowary (*Casuarius* spp.); the noisiest is probably the black, red-cheeked palm cockatoo (*Prosciger aterrimus*), whose huge hooked bill can crack even the hardest nuts of the forest. The squawks of the white, sulphur-crested cockatoo (*Cacatua galerita*) may also be heard in the lowland forests.

There are also nine megapode birds akin to the maleo of Sulawesi and related Australian species: the brush turkey (five species) and the scrubfowl (four species), which incubate their eggs in warm mounds of decaying leaf litter.

Like most lories, this Western black-capped lory (*Lorius lory*) feeds chiefly on nectar, pollen and soft fruit, and is brilliantly coloured. Lories will travel considerable distances to follow the seasonal flowering and fruiting patterns of their favourite food trees.

The rarely seen vulturine parrot, also known as Pesquet's parrot (*Psittrichas fulgidus*), is endemic to Irian Jaya and prefers montane forests betwen 800 and 2000 metres up. Its long beak, used to probe soft fruit, flowers, and honey, gives it a vulture-like appearance.

A raucous braying call announces the presence of the black palm cockatoo (*Probosciger aterrimus*) in the Irian Jaya forests. With its powerful bill, it can open extremely hard seeds and fruits, such as those of palms and pandans. The bird is usually solitary and nests in hollow trees.

A pair of white sulphur-crested cockatoos (*Cacatua galerita*), commonly found in the lowland forests of Irian Jaya.

Perhaps the strangest bird of all is the blackbird-like melampitta (*Melampitta* spp.), which lives practically underground, in deep limestone sinkholes often created by waterfalls, emerging only at dusk to feed on fruit and insects. As ecologist Kathy MacKinnon has aptly put it, the bird 'creeps about like a rat', although it is perfectly capable of flight.

Other animals range from the bizarre to the alarming. The toothless echidna or spiny anteaters resemble hedgehogs, rolling themselves up into a ball when annoyed or threatened, although if harassed on soft ground, they will burrow into it. Echidnas feed by tearing open termite nests and decaying logs with their sharp claws and gathering insects on their long sticky tongues. The short-beaked species can extend its tongue 18 centimetres from the tip of the snout. In the long-beaked species, the tip of the tongue has backward-pointing spines to help the echidna grip earthworms, the species' main food. Female echidnas lay only one egg a year, and carry it around in a pouch on the belly, where the baby suckles its mother's milk once it has broken out of the egg.

Among Irian Jaya's marsupials are 7 species of bandicoot (Peramelidae), 13 species of marsupial cat and mouse (Dasyuridae), 8 species of cuscus (Phalangeridae), 16 species of possum and ringtail (Burramyidae and Petauridae), including the sugar glider (*Petaurus breviceps*), 5 species of forest

wallaby (*Dorcopsis* spp.), and 3 species of tree kangaroo (*Dendrolagus* spp.). Almost every schoolchild is familiar with the unique features of marsupials, the most famous flagcarrier for the family being the kangaroo. The strength of the instinct that guides tiny new-born marsupials, be they possums or kangaroos, still looking like the embryos they virtually are, to find their mother's pouch and climb into it to suckle at a hidden teat, can only be marvelled at.

One endearingly furry marsupial, the tree-dwelling cuscus, is, unfortunately, all too often first seen as a dead adornment for human bodies in Irian Jaya. The native peoples also like to eat this bear-like creature, and they have six species to choose from in Irian Jaya. Carnivorous marsupial mice may be related to the supposedly extinct Tasmanian 'tiger' or wolf of Australia, while small marsupial cats occupy the ecological niche of the big cats in other countries.

Inland lakes are stocked with life. Massive, 5-metre-long small-toothed sawfish (*Pristiopsis leichardti*) patrol the larger lakes, feeding on other fish. A shark-like predator, the sawfish feeds on the dead and dying fish it has mauled with its jaws. Sawfish are well protected by some of the local people, who believe they are the living receptacles for their own ancestral spirits.

Among the more dangerous animals are the huge estuarine crocodiles (*Crocodylus porosus*), now forming the basis of an important, potentially sustainable crocodile-ranching industry, which exports the animals' skins. The largest animal to be found in Irian Jaya, this crocodile has been recorded at lengths as incredible as 7 metres. There is also the less well-known New Guinea crocodile (*Crocodylus novaeguineae*), its main Irian Jaya location being the mighty 800-kilometre Mamberamo River in the north.

The New Guinea crocodile (*Crocodylus novaeguineae*) is less commonly seen than the estuarine or salt-water crocodile (*Crocodylus porosus*), and lives mainly along the Mamberamo River to the north of Irian Jaya. Both species are exploited for their skins.

The New Guinea green tree python (*Chondropython viridis*) from Irian Jaya is non-venomous, and attractive enough to be in demand as a pet. The young of this species are quite differently coloured, variously brown, yellow, pink, or red (see page xv at the front of this book).

Also on the alarming side are the 800 spider species, which include the giant bird-eating spider (*Selencosmia crassides*), and more than 100 snakes, with the very aggressive death adder (*Acanthopsis antarcticus*), whose bite can bring death within minutes, topping the list, followed by the equally venomous taipan (*Oxyuranus scutellatus*).

In some uninhabited areas, however, the animals are so unused to Man they are said to be almost tame. Some of the most beautiful creatures are totally harmless, and indeed are themselves in some danger from collectors: the birdwing butterflies, with wing-spans up to 20 centimetres, are worth as much as US$1,000 each on the international market.

Nobody yet understands why some birdwing species' caterpillars destroy their host plant when they pupate, a gesture that seems suicidal in terms of the survival of the whole species. The latest moves to farm birdwings may help solve such puzzles, as study of these gorgeous creatures' food plants and life-cycles becomes a matter of practical necessity for human butterfly ranchers.

Other fascinating, if odd, insects include stick insects over 30 centimetres long, and antlered flies, the males adorned with elaborate horns.

Among the 56 protected areas in Irian Jaya, 12 of them marine (including 32 areas proposed but not gazetted), the chief galleries for all these creatures are the Mount Lorentz Reserve, at 21 500 square kilometres, Indonesia's largest national park, the 14 425-square-kilometre Mamberamo–Foja

The birdwings are beautifully coloured, giant butterflies and Irian Jaya is at the centre of their distribution. The one pictured here is the *Ornithoptera goliath*.

Boelen's python (*Python boeleni*) is the only python found at high altitudes, up to 3000 metres. It can grow to lengths of over 3 metres.

Reserve, the 3120-square-kilometre Wasur–Rawa Biru National Park, a monsoonal forest and wetlands area, and the 450-square-kilometre Arfak Reserve.

Extraordinary variety of habitat is the distinguishing feature of the Mount Lorentz Reserve located in Central Irian Jaya. It includes Indonesia's, and South-East Asia's, highest mountain—Puncak Jaya, at 5030 metres—and features the spectacular glaciers of the Carstensz Range. This is one of only three places in the world where glaciers exist at equatorial latitudes. Scientists have assessed 34 different vegetation types, including alpine types, within the reserve, a microcosm of Irian Jaya.

Mamberamo–Foja, as its name suggests, contains the province's largest river, the Mamberamo, with its countless ox-bow lakes, the province's largest lake, Lake Bira, 15 kilometres long, and the wild Foja Mountains area, where it is said tree kangaroos and cuscus, unafraid, allow humans to get very close.

The giant blue-tongue lizard (*Tiliqua gigas*), a skink, demonstrates the reason for its name.

The crocodile skink or helmeted skink (*Tribolonotus gracilis*) is one of Irian Jaya's more than 100 skink species.

The Wasur National Park, located in the far south-east of Irian Jaya, is distinctive for its wetlands, a sympathetic habitat for wading birds and unusual lizards like the endemic giant monitor lizard (*Varanus salvadorii*) and the frilled lizard (*Chlamydosaurus kingi*). There is also a healthy population of introduced rusa deer (*Cervus timorensis*).

The Arfak Reserve lies in the 'Bird's Head' peninsula at Irian Jaya's north-western end, and although an important centre for rare mountain birds, is most famous for the Rothschild's birdwing butterfly (*Ornithoptera rothschildi*) found nowhere else; like the crocodile, birdwings are now the basis for a potentially lucrative farming industry. The reserve also houses 30 marsupials, 21 of them endemic to New Guinea, and more than 150 New Guinea-endemic species of bird, not to mention 17 New Guinea-endemic rodents. Together with the Cyclops Nature Reserve, Arfak is a site

The idea is to alarm the enemy, but in truth this frilled lizard (*Chlamydosaurus kingi*) is not particularly harmful, except to the insects it eats. When the lizard is at rest, the frilled collar is folded back and almost invisible.

where there has been some success with the idea of involving local indigenous peoples in the management and conservation of a reserve. The World Wide Fund for Nature's experiences in Irian Jaya, in particular, have shown that the drafting of any successful conservation management plan now must include prior sociological studies of the local people's own traditional patterns of land-use and land-management.

Irian Jaya has been left wild for so long and for the moment has so little population pressure to contend with, that it offers a unique opportunity to demonstrate Man's respect for some of the world's most thrilling forests.

There are 16 species of possum and ringtail on New Guinea. Here is the common striped possum (*Dactylopsila trivirgata*), related to the cuscus and even the Australian koala.

Treetop Teddy Bears

'Phalanger' is the ponderous name with which a group of cuddly, furry marsupial animals is loosely burdened. In Australia, this group includes the world-famous koala (*not* a bear!), and the whole group is often generally referred to there as 'possums'. However, only two types of phalanger are found on Irian Jaya—possums and cuscus.

Some phalanger species look a little like mini-bears or monkeys, others more like cats or squirrels, still others like mice or rats, and their sizes vary accordingly. Phalangers have claws specially designed to comb through their deep woolly fur. They live mostly in the treetops and are largely vegetarian, eating fruit and flowers, with only the occasional insect for variety—although some species are more eclectic killers. Phalangers are mostly nocturnal, as their large, light-catching eyes suggest, and several have long prehensile, branch-gripping, tails.

Ecologist Kathy MacKinnon has remarked how phalangers seem to fill the forest niches usually occupied by primates and squirrels in Indonesia's western regions. The endearing little sugar glider (*Petaurus breviceps*) is one example, having taken on a role similar to that of the flying squirrel elsewhere. Here is a marsupial which has developed gliding as its chief mode of locomotion. A flap of skin between the sugar glider's limbs can be stretched out like wings, allowing it to leap into the void and glide freely for up to 50 metres. During the day, the sugar glider rests on a nest of leaves which it has hand-picked for itself, storing the day's haul in the coils of its tail.

Like the European dormouse, some possums can virtually suspend life for a while if the weather gets too cold, becoming literally 'dormant'. The cuscus, another slow mover, at first sight resembles the slow loris found in western Indonesia. Males and females of the same species often display dramatically different markings. The male spotted cuscus (*Spilocuscus maculatus*), for instance, is creamy white with some black patches, whereas the female is grey.

Unfortunately for the wide-eyed, slow-moving, and smelly cuscus, it makes good eating, and bracelets of cuscus fur look good on Irianese men's arms.

The pretty sugar glider (*Petaurus breviceps*) can indeed glide, for up to 50 metres between trees, thanks to flaps of skin between its limbs. This marsupial gathers leaves to build nests in hollow trees for its rest period during daylight hours.

Flying Roos

A kangaroo up a tree does not fit the imagination of most Western childhoods, but among the kangaroos of Irian Jaya are indeed tree kangaroos (*Dendrolagus* spp.), which are very good climbers. In fact, they much prefer to be up a tree than on the ground, where they look particularly awkward, hopping painfully with their tails held up straight. Their diet reflects their physical niche, consisting mainly of leaves, creepers, and fruits available up in the trees.

Naturally, tree kangaroos do not need the same powerful or disproportionately huge hind legs sported by their more terrestrial relatives in Australia. Their legs are more evenly matched, although their forepaws are very wide and equipped with rough foot-pads as well as large, strong claws for gripping on to branches and tree-trunks. Their sturdy tails are sometimes used as a rudder when they jump between trees, often across gaps 6 metres wide.

These animals have long been hunted by the native peoples of Irian Jaya and as a result have become very shy, making them difficult to spot in the forest. Even the most attractively marked species of tree kangaroo blend into the forest quite inconspicuously, and the animal has learned to sit very still. Of the three species found in Irian Jaya, the grizzled tree kangaroo (*Dendrolagus inustus*) is particularly vulnerable in this respect because it spends less time in trees than relatives like the black or Vogelkop tree kangaroo (*Dendrolagus ursinus*), whose face may be adorned with either white or reddish side ruffs, or the unicoloured tree kangaroo (*Dendrolagus dorianus*).

The black or Vogelkop tree kangaroo (*Dendrolagus ursinus*) is found only in the Bird's Head Peninsula area of Irian Jaya, as its German name suggests. A lowland forest animal, it does sometimes descend from the trees to collect plant material and even small animals for food. (Bruce Coleman Limited)

God's Birds

Birds of paradise are, quite simply, the most astounding
and ravishingly beautiful birds in the world.

(Zoologist and TV personality, David Attenborough)

As Wallace described it in *The Malay Archipelago*, early European traders in the Moluccas were often presented with the dried skins of birds so strange and beautiful as to excite the admiration even of those wealth-seeking rovers. Malay traders had already tagged these wondrous creatures *manuk dewata* or 'God's birds', while the Portuguese sailors called them *passaros de sol* or 'birds of the sun'.

Because the birds' corpses were often sold with the feet amputated to hide the damage perpetrated by the crude native trapping methods, many early European explorers thought the birds of paradise had no feet. This explains the scientific Latin name chosen in the eighteenth century for the largest species, the greater bird of paradise, by the father of taxonomy, the science of classification, Linnaeus—*Paradisaea apoda*, or 'the footless bird of paradise'.

Perhaps because their natural habitat was deep in the forest, well away from the coastal trading towns, and because they were never seen alive by foreigners, the birds acquired a mythical reputation; they were said to live permanently in the air, never alighting on the earth, homing always towards the sun, equipped with neither feet nor wings, and feeding on dew and air. They dwelled, quite literally, in paradise. Wallace was the first visitor to report accurately on these birds, pronouncing them 'one of the most beautiful and most wonderful of living things'. By the end of the

'Lesser' seems a curiously deflating name for this handsome lesser bird of paradise (*Paradisaea minor*). Early explorers thought the feathers of such birds came from angels in Paradise.

Wilson's bird of paradise (*Cicinnurus respublica*) seen from the front while displaying its chest feathers.

▷
Wilson's bird of paradise is a small species. It is not easy to remember that these birds are closely related to crows and starlings.

nineteenth century, the famous publisher of bird books, John Gould, had issued paintings of more than 20 species. We know today that there are 42 species in the whole family. A few of them are found only in Australia or in the Moluccas; just 4 of them are endemic to Irian Jaya but 22 are found in both Papua New Guinea and Irian Jaya. We also know now that these gorgeous birds are most closely related to crows, not to angels as it was once thought.

Birds of paradise have been able to evolve without fear of major predators, hence the startling colours and seeming impracticalities of their extravagant plumage: fans, wires, shields, and streamers of elongated or strangely distorted feathers, painted emerald green, bright sulphur yellow or glittering gold, cobalt blue, deep ruby-red. The beauty of their plumage has not gone unnoticed among the tribal peoples of Irian Jaya, who value it as sexually attractive decoration for men and as ritual bargaining currency in negotiations such as fixing the bride-price. Wallace was made to understand the local value of the birds' plumage to his own cost. He was unable to obtain specimens of more than five species, because of lack of co-operation from the local people who guarded the birds as their own private treasure.

Viewing a group of about 500 plumed tribal dancers, famous naturalist David Attenborough recently estimated that the men must have slaughtered at least 10,000 birds of paradise to account for their exotic head-dresses. Bird of paradise feathers were similarly used by the be-hatted fashionable ladies of Europe in the nineteenth century, until a trade ban was imposed in the 1920s. It was this trade, above all, that triggered a decline in the numbers of certain more attractive species. Those species which display in groups are particularly vulnerable to hunting.

These birds are as remarkable for their behaviour as for their looks, in particular the way the dandyish, polygamous males of several species display to the exceedingly drab brown females. Wallace described them as being 'in constant motion'. Certain species' penchant for 'dancing' on the ground is another indication of their freedom from predators. While the male greater bird of paradise is among the species that courts in groups, making a great deal of noise and dancing around a lot, the male twelve-wired bird of paradise (*Seleucidis melanoleuca*) relies more on his dashing

The king bird of paradise (*Cicinnurus regius*) is another small species. However, unlike most of its fellows, this species nests in tree-holes.

The king bird of paradise seen from the front.

This is the magnificent riflebird (*Ptiloris magnificus*), which dances ecstatically with wings outspread and head bent back to display its metallic blue breast shield when courting.

The female bird of paradise looks quite different from the male and is often much duller. This female magnificent riflebird is having her turn at a bit of display during courting. But after mating, she will be left completely alone by the male to build her nest and tend her young.

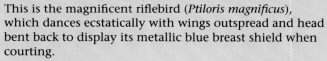

The red bird of paradise (*Paradisaea rubra*) can fan out its red tail feathers to magnificent effect, impressing the female. Only males sport the gorgeous plumage associated with these birds.

black-and-yellow plumage and the startling 12 long feathers curling from its tail like a fan which it flutters during courtship. Other species may buzz, as does the magnificent bird of paradise (*Cicinnurus magnificus*) or bow and hiss, like the King of Saxony bird of paradise (*Pteridophora alberti*); some sicklebills (*Epimachus* spp.) construct a screen of feathers around their heads through which they peer, while the superb bird of paradise (*Lophorina superba*) flings a feathery black cape around its head at the same time as displaying a green triangular shield on its chest.

As with the bowerbird, most male birds of paradise take no part in the rearing of young, leaving this entirely to the female. (Curiously, the scientific term for such males is 'emancipated'.) However, there are some monogamous birds of paradise which mate for life with one female and as a result need far less interesting plumage and behaviour.

Many of these birds are still not well known; naturalists still do not know where to find their nests, nor what they look like, for example. The birds apparently are omnivorous, eating anything from fruits to insects. Wallace managed to feed two captive lesser birds of paradise (*Paradisaea minor*) on a long sea voyage to England by hunting for shipboard cockroaches and loading bananas at ports of call. He reported that they withstood cold very well, and after arrival at London's Zoological Gardens, survived another two years. What the birds really needed to thrive, Wallace said, was 'air and exercise', or to use another word, freedom.

Beware the Cassowary's Kick

Zoologists call them 'ratites'. Most of us know them as 'flightless birds'. The cassowary of New Guinea falls into this group, as do the better known examples of the genre like the ostrich and the kiwi. But the blue-headed, red-wattled cassowary and the kiwi are the only ones of the group which live in forest rather than on open plains.

Without any large predator to fear, the cassowary has abandoned flight and, indeed, is too heavy to take off anyway, at about 55 kilograms, and standing as tall as a man, around 1.5–1.8 metres. There are three species of cassowary in Irian Jaya: the large double-wattled kind of the lowland forest and savannah called the southern cassowary (*Casuarius casuarius*), the dwarf species or Bennett's cassowary (*Casuarius bennetti*), from the mountain forests, and the single-wattled or northern cassowary (*Casuarius unappendiculatus*) found in swamps by rivers and along the northern coast.

The cassowary's strange bony casque protects the bird from thick undergrowth. Its 'feathers' are more like bristles than true feathers. The bird may look clumsy, but it can run at 40 kilometres per hour and hurdle considerable obstacles at one jump. A Thai-style kick-fight with a cassowary would be a losing battle: the innermost of its three toes bears a long sharp claw, with the potential for inflicting horrific injuries—usually complete disembowelment—on any foe, human or bird, when the bird leaps feet first at its target. It is the job of the powerful male to incubate and guard the 3–6 green eggs laid by the female in a ground-level nest, a job he pursues with some vigour. Despite these fearsome traits, the cassowary is often domesticated and kept as a pet in Irianese villages, at least until it becomes a more unpredictable adult.

The bird patrols the forest feeding mainly on fallen palm seeds and fruit, as well as fungi, insects, and small dead animals, occasionally giving a deep booming call or a softer croak. Its diet ensures wide dispersal of certain forest fruits and seeds, making the bird's survival important to the maintenance of the forest as a whole, yet another example of ecological networking.

The huge cassowary, here the double-wattled cassowary (*Casuarius casuarius*), may stand 1.5 metres high. Flightless, it can manage running speeds of up to 40 kilometres per hour. For defence, the bird is armed with a vicious claw on each of its feet.

Epilogue

I

T has been a recurring theme of this book that Man himself is part of the rain forest ecology. Human diversity is just as exciting and worthy of conservation as that of other species on this planet. Human survival is interlinked with the survival of all other species.

The theme is worth restating. We could do little better than turn to Wallace's parting words in *The Malay Archipelago*, written long before a United Nations 'Year of Indigenous People' was declared for 1993:

We most of us believe that we, the higher races, have progressed and are progressing. If so, there must be some state of perfection, some ultimate goal, which we may never reach, but to which all true progress must bring us nearer. What is this ideally perfect social state toward which mankind ever has been, and still is tending?...

Now it is very remarkable, that among people in a very low stage of civilization, we find some approach to such a perfect social state. I have lived with communities of savages in South America and in the East, who have no laws or law courts but the public opinion of the village freely expressed. Each man scrupulously respects the rights of his fellow, and any infraction of those rights rarely or never takes place. In such a community, all are nearly equal....

Now although we have progressed vastly beyond the savage state in intellectual achievements, we have not advanced equally in morals....

... Our mastery over the forces of nature has led to a rapid growth of population, and a vast accumulation of wealth; but these have brought with them such an amount of poverty and crime ... that it may well be questioned ... whether the evil has not overbalanced the good. Compared with our wondrous progress in physical science and its practical applications, our system of government, of administering justice, of national education, and our whole social and moral organization, remains in a state of barbarism. And if we continue to devote our chief energies to the utilizing of our knowledge of the laws of nature with the view of still further extending our commerce and our wealth, the evils which necessarily accompany these when too eagerly pursued, may increase to such gigantic dimensions as to be beyond our power to alleviate.

We should now clearly recognize the fact, that the wealth and knowledge and culture of *the few* do not constitute civilization, and do not of themselves advance us toward the 'perfect social state'....

This is the lesson I have been taught by my observations of uncivilized man.

The 'Noble Savage' ideology and paternalism that underlay such nineteenth-century tracts was flawed but still, Wallace's words ring across more than a century now with ever-increasing relevance to our own times: our dominion over Nature has not necessarily brought us happiness.

Only when we human beings realize that we are rooted in the natural world, not separate from it, will we achieve balance in our lives, and a secure future.

◁
Asmat tribesmen of Irian Jaya pay tribute to their own 'Tree of Life' at a sago tree ceremony—so-called 'primitive' peoples are more aware of their debt to Nature than so-called 'civilized' Man.

Bibliography

Achmadi, Heri, 'Raging Fire Wiping Out Tropical Forest', *Business Times* (Singapore), 6 November 1989.

Asiaweek, 'Wound in the World', Hong Kong, 13 July 1984.

Attenborough, David, 'Tales from Paradise', *BBC Wildlife*, Vol. 9, No. 9, September 1991.

Bernard, Hans-Ulrich, *S.E. Asia Wildlife*, Singapore: Apa Publications Insight Guides, 1991.

Buck, Frank, *On Jungle Trails*, London: George G. Harrap & Co. Ltd., 1939.

Cater, Bill, 'The Orang Gang', *BBC Wildlife*, Vol. 8, No. 8, August 1990.

Choo-Toh, Get Ten et al., *A Guide to the Bukit Timah Nature Reserve*, Singapore: Singapore Science Centre, 1985.

Covarrubias, Miguel, *Island of Bali*, New York: Alfred A. Knopf Inc., 1937; reprinted Kuala Lumpur: Oxford University Press, 1972.

De Jonge, Nico et al. (eds.), *Indonesia in Focus*, London: Kegan Paul International Ltd., 1991 (first published in Dutch, Netherlands, Edu' Actief Publishing Company, 1990).

Dekker, Rene W. R. J., *Conservation and Biology of Megapodes*, Amsterdam: Institute of Taxonomic Zoology, Amsterdam University, 1990.

De Wit, Augusta, *Java: Facts and Fancies*, The Hague: W. P. van Stockum, 1912; reprinted Kuala Lumpur: Oxford University Press, 1984.

Diamond, Jared, *The Rise and Fall of the Third Chimpanzee*, London: Radius, 1991.

Directorate-General of Tourism, Indonesia, leaflets and brochures.

Encyclopaedia Britannica.

Far Eastern Economic Review Asia Yearbook, 1990.

Henderson, M. R., *Common Malayan Wildflowers*, Kuala Lumpur: Longman Malaysia, 1961.

Holmes, Derek and Nash, Stephen, *The Birds of Sumatra and Kalimantan*, Singapore: Oxford University Press, 1990.

Holttum, R. E., *Plant Life in Malaya*, Kuala Lumpur: Longman Malaysia, 1954.

Hose, Charles, *The Field-Book of a Jungle-Wallah*, London, H. F. & G. Witherby, 1929; reprinted Singapore: Oxford University Press, 1983.

Johnson, Dennis, 'Palms in Asia: The Broader View', in Dennis Johnson (ed.), *Palms for Human Needs in Asia: Palm Utilization and Conservation in India, Indonesia, Malaysia and the Philippines*, n.p.: Balkema, 1991.

Kiew, Ruth and Davison, G. W. H., 'Relationship between Wild Palms and Other Plants and Animals', in Dennis Johnson (ed.), *Palms for Human Needs in Asia: Palm Utilization and Conservation in India, Indonesia, Malaysia and the Philippines*, n.p.: Balkema, 1991, Appendix I.

Latham, R. E. (trans.), *The Travels of Marco Polo*, Harmondsworth: Penguin Books, 1958.

Loeb, Edwin M., *Sumatra: Its History and People*, Vienna: Verlag des Institut fur Volkerkunde der Universitat Wien, 1935; reprinted Kuala Lumpur: Oxford University Press, 1972.

Lulofs, Madelon H., *Coolie*, London: Cassell & Co. Ltd., 1936; reprinted Kuala Lumpur: Oxford University Press, 1982.

McDougal, Charles, *The Face of the Tiger*, London: Rivington Books in association with Andre Deutsch, 1977.

McKie, Ronald, *The Company of Animals*, Australia: Angus & Robertson Ltd.; reprinted Melbourne: Readers Book Club, 1967.

MacKinnon, John, *The Ape Within Us*, London: William Collins Sons & Co. Ltd., 1978.

MacKinnon, Kathy, *Nature's Treasurehouse: The Wildlife of Indonesia*, Jakarta: Penerbit PT Gramedia Pustaka Utama, 1992.

McNeely, Jeffrey A. and Wachtel, Paul Spencer, *Soul of the Tiger: Searching for Nature's Answers in Exotic Southeast Asia*, New York: Doubleday, 1988; reprinted Singapore: Oxford University Press, 1991.

Malayan Nature Society, *Endau–Rompin: A Malaysian Heritage*, Kuala Lumpur: Malayan Nature Society, 1988.

Maxwell, George, *In Malay Forests*, Edinburgh and London: William Blackwood & Sons, Ltd., 1907.

Mjöberg, Eric, *Forest Life and Adventures in the Malay Archipelago*, London: George Allen & Unwin Ltd., 1930; reprinted Singapore: Oxford University Press, 1988 (first published in Swedish, 1928).

Mogea, Johanis P., 'Indonesia: Palm Utilization and Conservation', in Dennis Johnson (ed.), *Palms for Human Needs in Asia: Palm Utilization and Conservation in India, Indonesia, Malaysia and the Philippines*, n.p.: Balkema, 1991.

Muller, Kal, *Indonesian New Guinea: Irian Jaya*, Singapore: Periplus Editions Indonesia Travel Guides, 1991.

_____, *Kalimantan: Indonesian Borneo*, Singapore: Periplus Editions Indonesia Travel Guides, 1990.

_____, *Maluku: The Moluccas*, Singapore: Periplus Editions Indonesia Travel Guides, 1991.

Oey, Eric M., *Java*, Singapore: Periplus Editions Indonesia Travel Guides, 1991.

_____, *Sumatra*, Singapore: Periplus Editions Indonesia Travel Guides, 1991.

Paddoch, Christine et al., 'Complexity and Conservation of Medicinal Plants: Anthropological Cases from Peru and Indonesia', in *The Conservation of Medicinal Plants: Proceedings of an International Consultation*, 21–27 March 1988, WHO/IUCN/WWF, in Chiang Mai, Thailand, Cambridge: Cambridge University Press.

Passmore, Jacki, *The Encyclopedia of Asian Food and Cooking*, n.p.: Doubleday, 1991.

Payne, Junaidi and Andau, Mahedi, *Orang-Utan: Malaysia's Mascot*, Kuala Lumpur: Berita Publishing, 1989.

Petocz, Ronald G., *Conservation and Development in Irian Jaya: A Strategy for Rational Resource Utilization*, Leiden: E. J. Brill, 1989.

Polunin, Ivan, *Plants and Flowers of Malaysia*, Singapore: Times Editions, 1988.

_____, *Plants and Flowers of Singapore*, Singapore: Times Editions, 1987.

Reuter, 'Fire Destroys 30,000 ha Forest in Indon Borneo', *New Straits Times* (Malaysia), 1 October 1991.

Rifai, Mien A. and Kartawinata, Kuswata, 'Germplasm, Genetic Erosion and the Conservation of Indonesian Medicinal Plants', in *The Conservation of Medicinal Plants: Proceedings of an International Consultation*, 21–27 March 1988, WHO/IUCN/WWF, in Chiang Mai, Thailand, Cambridge: Cambridge University Press.

Rubeli, Ken, *Tropical Rain Forest in South-East Asia: A Pictorial Journey*, Kuala Lumpur: Tropical Press, 1986.

Shah, Idries, *Oriental Magic*, first published in Great Britain by Rider & Co, 1956; reprinted St. Albans: Paladin, 1973.

Shelford, Robert W., *A Naturalist in Borneo*, London: T. Fisher Unwin Ltd., 1916; reprinted Singapore: Oxford University Press, 1985.

Sullivan, Francis, 'Rebirth in Bali', *BBC Wildlife*, Vol. 8, No. 8, August 1990.

Sutarjadi; Hakim, Lukman; and Ilman, Amirul, 'Medicinal Plants of East-Javan Forests: Prospects and Problems in Production and Utilisation', in Khozirah Shaari, Abd. Kadir Azizol, and Mohd. Ali Abd. Razak (eds.), *Medicinal Products from Tropical Rain Forests: Proceedings of the Conference*, Kuala Lumpur: Forest Research Institute of Malaysia, 1992.

Traffic Bulletin, Traffic International, Cambridge, 1980–93.

Veevers-Carter, W., *A Garden of Eden: Plant Life in South-East Asia*, Singapore: Oxford University Press, 1986.

_____, *Riches of the Rain Forest*, Singapore: Oxford University Press, 1984; reprinted, 1991.

Volkman, Toby Alice and Caldwell, Ian (eds.), *Sulawesi: The Celebes*, Singapore: Periplus Editions Indonesia Travel Guides, 1992.

Wallace, Alfred Russel, *The Malay Archipelago*, London: Macmillan, 1869; reprinted Singapore: Graham Brash, 1983, and Singapore: Oxford University Press, 1986.

Whitten, Anthony J.; Mustafa, Muslimin; and Henderson, Gregory S., *The Ecology of Sulawesi*, Jakarta: Gadjah Mada University Press, 1987.

Whitten, Anthony J. et al., *The Ecology of Sumatra*, Jakarta: Gadjah Mada University Press, 1987.

Whitten, Tony and Whitten, Jane, *Wild Indonesia: The Wildlife and Scenery of the Indonesian Archipelago*, London: New Holland (Publishers) Ltd., in association with the World Wide Fund for Nature, 1992.

WWF Indonesia, *Conservation Indonesia*, Jakarta, 1991–2.

Wurtzburg, C. E., *Raffles of the Eastern Isles*, London: Hodder & Stoughton, 1954; reprinted Singapore: Oxford University Press, 1984.

Yayasan Indonesia Hijau (Green Indonesia Foundation), *Voice of Nature (Suara Alam)* monthly, No. 34, October 1985; No. 36, December 1985.

Yong Hoi-Sen, *Malaysian Butterflies: An Introduction*, Kuala Lumpur: Tropical Press, 1983.

Zantovska, Jana, *Plants in Danger: What Do We Know?*, n.p., 1986.

Publications from the International Union for the Conservation of Nature and Natural Resources (IUCN), Gland and the World Wide Fund for Nature International (WWF International), Gland.

IUCN

Bachruddin, Memet A.; Irving, Alan; and MacKinnon, Kathy, 'Local Industry Support for Kutai National Park', Paper presented at World Parks Congress, Caracas, February 1991.

Ba Kader et al., 'Basic Paper on the Islamic Principles for the Conservation of the Natural Environment: An Islamic Study', IUCN/Meteorological and Environmental Protection Administration (MEPA) of the Kingdom of Saudi Arabia, 1983.

De Beer, Jenne H. and McDermott, Melanie, 'The Economic Value of Non-timber Forest Products in South-East Asia, with Emphasis on Indonesia, Malaysia and Thailand', Netherlands Committee IUCN, 1989.

Holdgate, Martin, 'Managing the Future', IUCN 40th anniversary lecture, London, 1989.

Janis, Ramon and Suratri, Retno, 'Siberut: Conserving Biodiversity and Traditional Cultures', Paper presented at World Parks Congress, Caracas, February 1991.

Johns, Andrew D., 'Species Conservation in Managed Forests', Paper presented at IUCN 18th general assembly, Wildlife Conservation International/New York Zoological Society, 1990.

MacKinnon, John and MacKinnon, Kathy, 'Review of the Protected Areas System in the Indo-Malayan Realm', IUCN/UNEP, 1986.

MacKinnon, John et al., 'Managing Protected Areas in the Tropics', 1986.

Manuwoto and Soemarsono, 'Kerinci–Seblat National Park: Focus for an Integrated Conservation and Development Project', Paper presented at World Parks Congress, Caracas, February 1991.

Mulyana, Yaya and Susetyo, Hart Lamer, 'A Case Study in Bromo Tengger Semeru and Baluran National Park', Paper presented in Tourism and Protected Areas session, World Parks Congress, Caracas, February 1991.

Ramono, Widodo S.; Santiapillai, Charles; and Sudarmadji, 'The Javan Rhino in Ujung Kulon National Park', Paper presented in Managing Endangered Species in Protected Areas session, World Parks Congress, Caracas, February 1991.

Rifai, Mien A., 'Useful Plants in Indonesian Reserves', Paper presented in Managing Genes and Species session, World Parks Congress, Caracas, February 1991.

Santiapillai, Charles and Jackson, Peter, 'The Asian Elephant: An Action Plan for Its Conservation', 1990.

Sayer, Jeffrey, 'Rainforest Buffer Zones: Guidelines for Protected Area Managers', 1991.

Setyowati-Indarto, N.; Wirjoatmodjo, S.; and Nasution, Rusdy E., 'Wild Fruit Trees of Economic Value in Indonesian Protected Areas', Paper presented in Managing Genes and Species session, World Parks Congress, Caracas, February 1991.

Siswanto, Wandojo and Marsh, Bruce E., 'The Gunung Lorentz Nature Reserve, Indonesia—Cooperative Developments between the Government, Mining Interests and Environmental Conservation', Paper presented at World Parks Congress, Caracas, February 1991.

Sumardja, Effendy A. and Siallagan, Toga, 'Transfrontier Reserves in Borneo and Asean', Paper presented in Management Challenges in Transfrontier Reserves session, World Parks Congress, Caracas, February 1991.

Triwibowo and Subagiadi, Herry, 'Building Public Support for Protected Areas: A Case Study in Meru Betiri National Park', Paper presented in Social Perception and Protected Areas session, World Parks Congress, Caracas, February 1991.

Wirjoatmojo, Soetikno; Susilo, Herry D.; and Sumin, 'Conservation in Mountain Habitats: Indonesian Situation—Java', Paper presented in Conservation in Mountain Habitats session, World Parks Congress, Caracas, February 1991.

WWF International

Elliott, Chris, *Tropical Forest Conservation*, WWF International Position Paper No. 7, September 1991.

Hails, Chris (comp.), *The Importance of Biological Diversity: A Statement by WWF*, c.1990.

Indonesia: Country Profile, May 1992.

Irian Jaya Fact Sheet, July 1992.

Markham, Adam (ed.), *Asian Tropical Forests*, Special Report, 1990.

Martin, Claude, *Panda*, No. III, August 1980.

'The Town of Sweet Wood'.

The Wild Supermarket: The Importance of Biological Diversity to Food Security, 1986.

Wachtel, Paul, 'Butterflies in the "Bird's Head" Might Save Nature Reserve and Benefit Local People'.

Index